Computer Simulati

MW01706754

to accompany

Electricity for the Trades

Frank D. Petruzella

Mc Graw Hill **Higher Education**

Boston Burr Ridge, IL Dubuque, IA Madison, WI New York San Francisco St. Louis
Bangkok Bogotá Caracas Kuala Lumpur Lisbon London Madrid Mexico City
Milan Montreal New Delhi Santiago Seoul Singapore Sydney Taipei Toronto

The McGraw-Hill Companies

COMPUTER SIMULATION LAB MANUAL TO ACCOMPANY ELECTRICITY FOR THE TRADES
FRANK D. PETRUZELLA

Published by McGraw-Hill Higher Education, an imprint of The McGraw-Hill Companies, Inc., 1221 Avenue of the Americas, New York, NY 10020.

This book is printed on acid-free paper.

1 2 3 4 5 6 7 8 9 0 DCD/DCD 0 9 8 7 6 5

ISBN 0-07-351082-3

www.mhhe.com

CONTENTS

Preface

The Computer Simulation Lab Manual for "Electricity for the Trades". Using Electronics Workbench/Multisim has been designed to reinforce the theory learned in basic DC and AC courses. This lab manual, designed for use with Electronics Workbench Multisim 2001 (Version 6 or higher) provides an easy and effective way to carry out circuit analysis.

Students work with circuits constructed on the computer screen and take measurements using simulated electrical instruments that act like the real thing. The lab assignments may require students to record measured data, calculate expected values, troubleshoot circuits, and answer questions designed to reinforce the theory associated with each experiment. Students run the simulations and compare the simulated results with that of the theoretical calculations and are required to answer questions to analyze the results.

The CD-ROM packaged with the lab manual contains some 250 practical circuits in files that are required to be loaded individually into the Multisim program for each lab assignment. The Electronics Workbench software itself is ***not*** included.

While nothing can replace the "hands-on" experience of actually experimenting with a real circuit, computer run lab experiments have a number of advantages.

- They are very cost effective compared to hardwired experiments.
- The availability of computers versus hardware components allows students to work individually on assignments instead of in pairs.
- Provides a safe alternative for experimenting with high-voltage circuits.
- Several circuits may be simulated in the time it takes to hardwire one equivalent circuit.
- Many of the simulation exercises can serve as a pre-lab for the hands-on lab.
- Circuit setups can be saved on disk.

Frank Petruzella

Acknowledgements

The author and publisher gratefully thank the following people for their careful analysis of each lab:

Alex Gregersen
Alameda County Electrical JATC
San Leandro, CA

Mel Switzer
Alameda County Electrical JATC
San Leandro, CA

As this lab manual was designed to accompany "Electricity for the Trades," the author and publisher would also like to acknowledge the following individuals who reviewed that textbook:

Don Kruckenberg
North Dakota State College of Science
Wahpeton, ND

Carl Latona
Santa Clara County JATC
San Jose, CA

Leland Peich
Blackhawk Technical College
Janesville, WI

Clyde Perry
GateWay Community College
Phoenix, AZ

Max Saravia
Houston Community College
Houston, TX

Ernest Shaffer
San Diego Electrical Training Trust
San Diego, CA

Mike Turner
Blackhawk Technical College
Evansville, WI

Andy Volper
San Diego JATC
San Diego, CA

Jim Westfall
San Diego JATC
San Diego, CA

NAME ___________________________ DATE _______________

UNIT 1

SAFETY

Multisim Simulation Assignments

1-1 Open the Multisim circuit file CH01_01 shown.

(a) Calculate the body current flow [in amperes (A) and milliamperes (mA)] of an electric shock victim who comes in direct contact with a 120-volt energy source. Assume a body contact resistance of 1000 ohms.

(b) Complete the schematic diagram with the 120-volt source, ammeter, and 1000-ohm resistor connected in series.

(c) Simulate the operation of this circuit and record the value of the current flow.

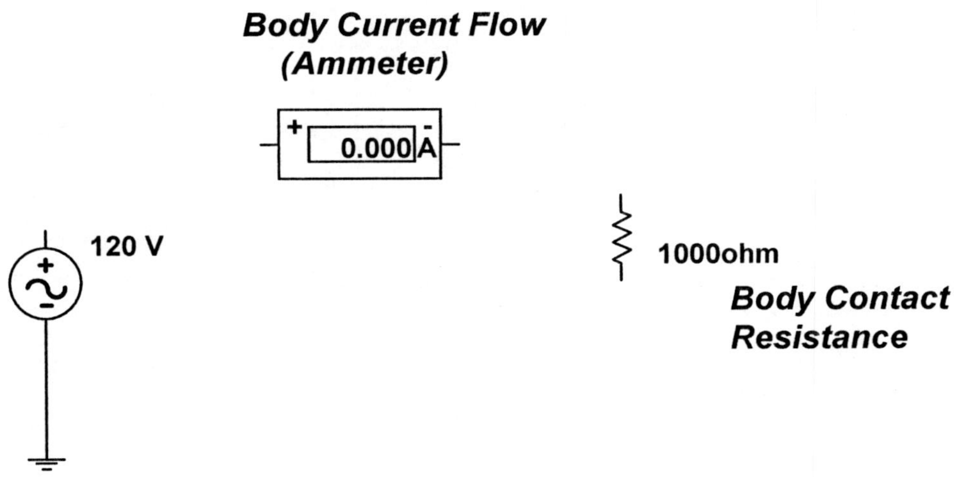

Answers

NAME ________________________________ DATE ________________

UNIT 2

INSTRUMENTS

Multisim Simulation Assignments

2-1 Open the Multisim circuit file CH02_01 shown.
A single multimeter can be used to measure voltage, current, or resistance.
(a) What electrical quantity is being measured by multimeter #1 and #2?
(b) Switch the simulation "on". Close switch S and record the value of the electrical quantity being measured by multimeter #1 and #2 (double click on the meter to view its control panel).

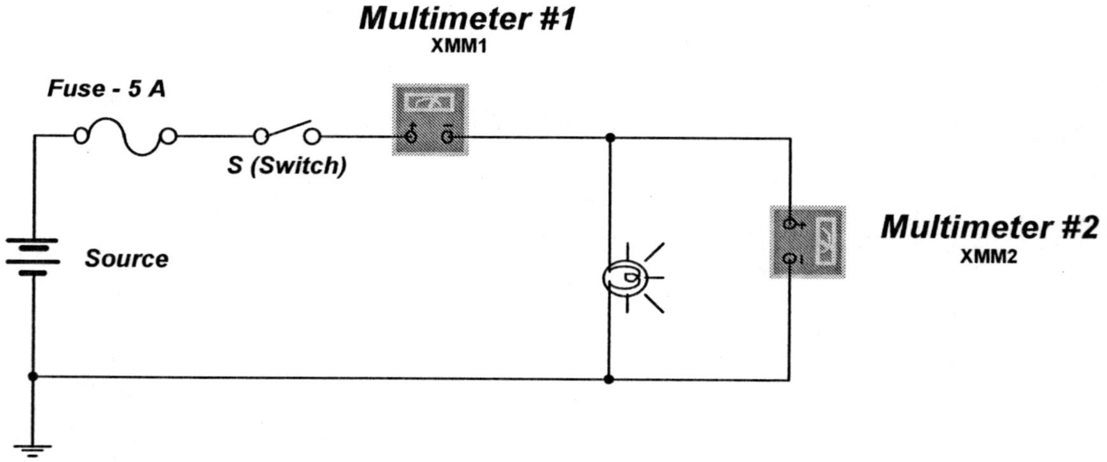

Answers

NAME ________________________________ DATE ________________

UNIT 3

CONDUCTORS, SEMICONDUCTORS, AND INSULATORS

Multisim Simulation Assignments

3-1 Open the Multisim circuit file CH03_01 shown.
The continuity tester shown can be used to test a fuse out-of-circuit for electrical continuity. Switch the simulation "on" and connect the continuity tester to test the condition of each of the four fuses (F1, F2, F3, F4) to determine which, if any, are blown.

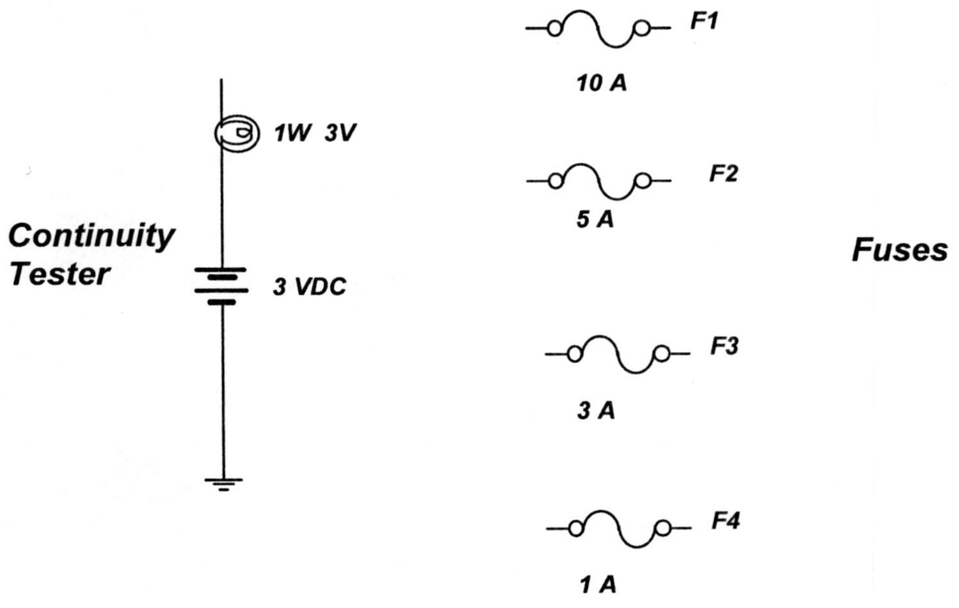

<u>Answers</u>

NAME ______________________________ DATE __________________

UNIT 4

SOURCES AND CHARACTERISTICS OF ELECTRICITY

Multisim Simulation Assignments

4-1 Open the Multisim circuit file CH04_01 shown.

Current electricity may be classified as direct current (DC) or alternating current (AC) based upon the voltage source.

(a) Switch the simulation "on". Close key A switch, and by examination of the oscilloscope waveform (double click on the scope to view the waveform) determine if the voltage source A is AC or DC.

(b) Open key A switch and close key B switch. By examination of the oscilloscope waveform determine if the voltage source B is AC or DC.

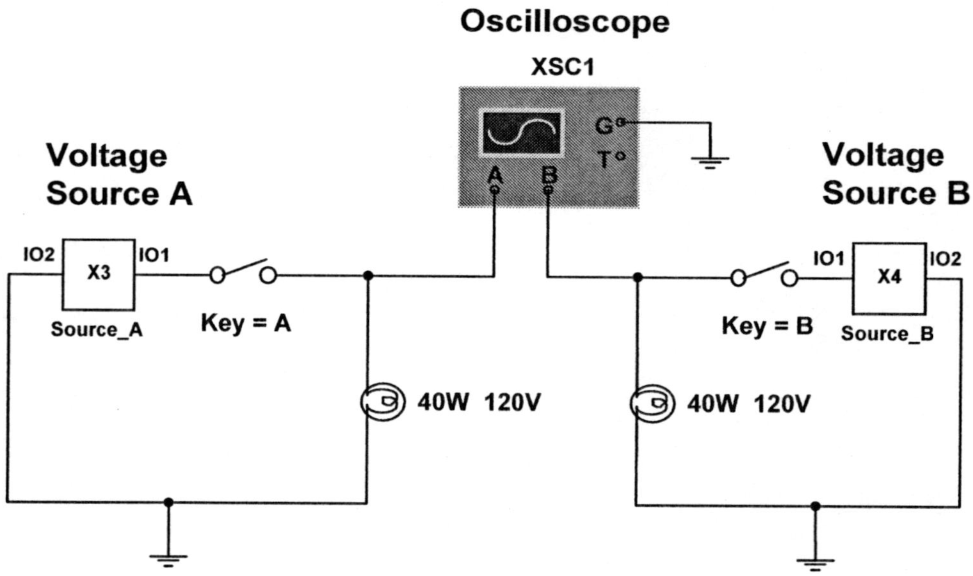

Answers

NAME ______________________________ DATE ______________

UNIT 5

BASIC ELECTRICAL UNITS

Multisim Simulation Assignments

5-1 Open the Multisim circuit file CH05_01 shown.
(a) Complete the schematic diagram properly connecting the battery to the lamp and ammeter.
(b) Simulate the operation of this circuit and record the amount of current drawn by the lamp.

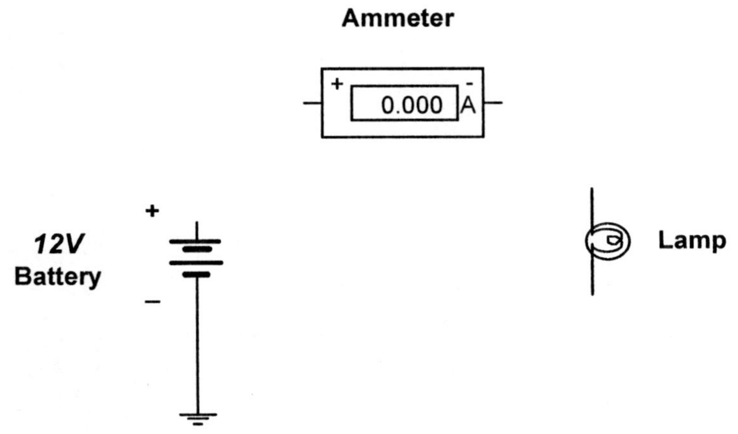

Answers

NAME ______________________ DATE ______________

5-2 Open the Multisim circuit file CH05_02 shown.

(a) Complete the schematic diagram properly connecting voltmeter #1 to measure the battery voltage and voltmeter #2 to measure the voltage of the lamp.

(b) Simulate the operation of this circuit and record the value of the battery voltage and the voltage of the lamp.

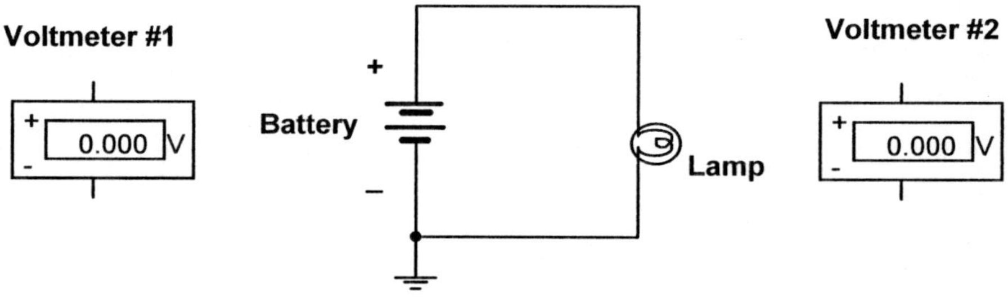

Answers

5-3 Open the Multisim circuit file CH05_03 shown.
Switch the simulation "on" and use the ohmmeter to measure the resistance value of R1, R2, and R3.

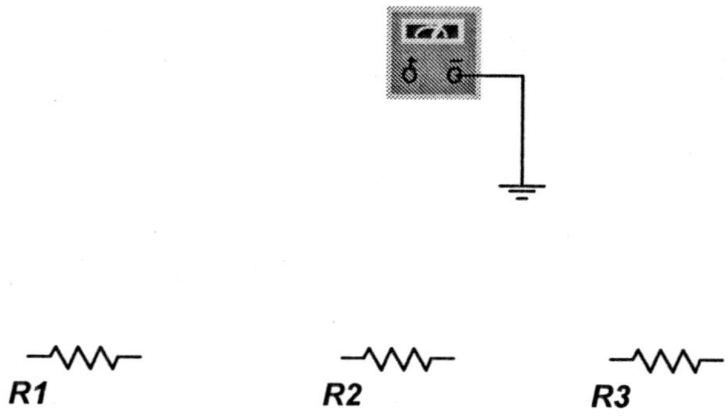

Answers

NAME ________________________________ DATE ________________

5-4 Open the Multisim circuit file CH05_04 shown.

(a) Turn the simulation "on" and apply 5V to the 5Ω resistor and record the current flow.

(b) Apply 10V to the 5Ω resistor and record the current flow.

(c) Which circuit produced the most current? Why?

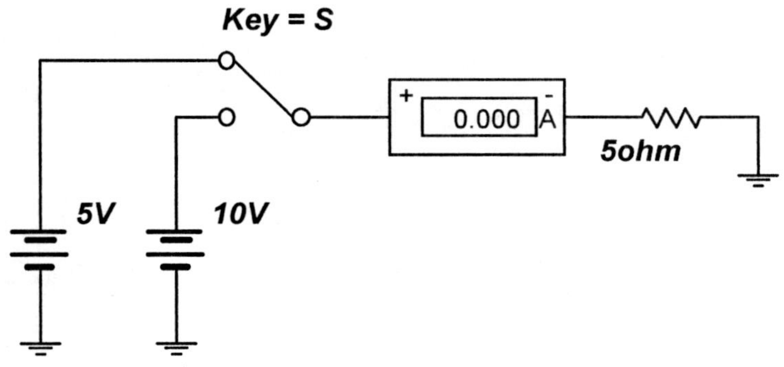

Answers

NAME ______________________________ DATE ______________

5-5 Open the Multisim circuit file CH05_05 shown.
A lamp draws a current of 0.5 A when 120 V is applied to it.

(a) Calculate the power rating of the lamp.
(b) Turn the simulation "on" and record the value of the voltage, current, and power.

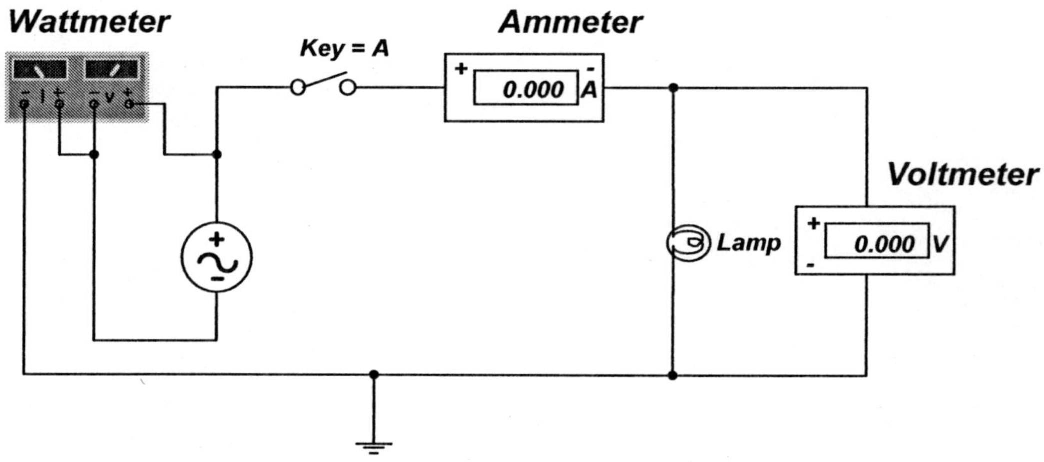

Answers

NAME ______________________________ DATE ________________

5-6 Open the Multisim circuit file CH05_06 shown.

(a) Using Ohm's law calculate the value of the current flow with 120 volts applied to the 10 Ω resistor.
(b) Turn the simulation "on" and record the value of the current.
(c) Using Ohm's law calculate the value of the current flow if the source voltage is decreased to 60 volts.
(d) Turn the simulation "on" and record the value of the current with a source voltage of 60 volts applied.
(e) Using Ohm's law calculate the value of the current flow if the source voltage is increased to 240 volts.
(f) Turn the simulation "on" and record the value of the current with a source voltage of 240 volts applied.

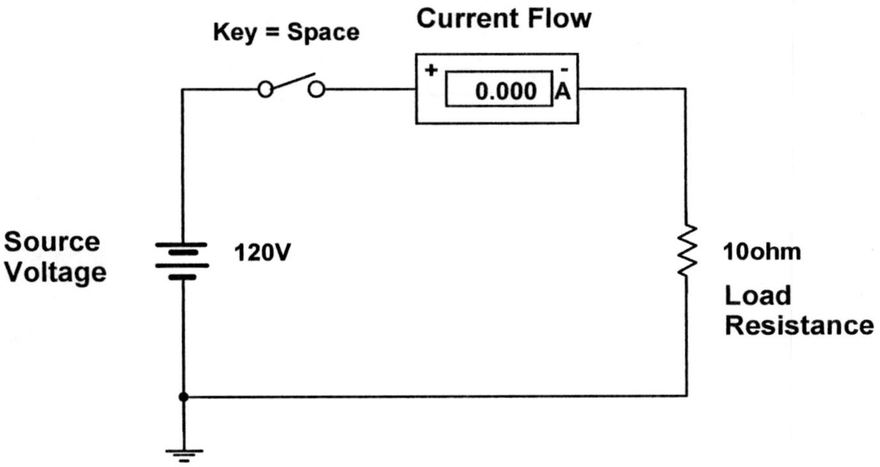

Answers

NAME ______________________________ DATE ____________

5-7 Open the Multisim circuit file CH05_07 shown.

(a) Using Ohm's law calculate the value of the current flow with 24 volts applied to the 4 Ω resistor.

(b) Turn the simulation "on" and record the value of the current.

(c) Using Ohm's law calculate the value of the current flow if the load resistance is decreased to 2 Ω (same voltage).

(d) Turn the simulation "on" and record the value of the current with a load resistance of 2 Ω connected.

(e) Using Ohm's law calculate the value of the current flow if the load resistance is increased to 8 Ω (same voltage).

(f) Turn the simulation "on" and record the value of the current with a load resistance of 8 Ω connected (same voltage).

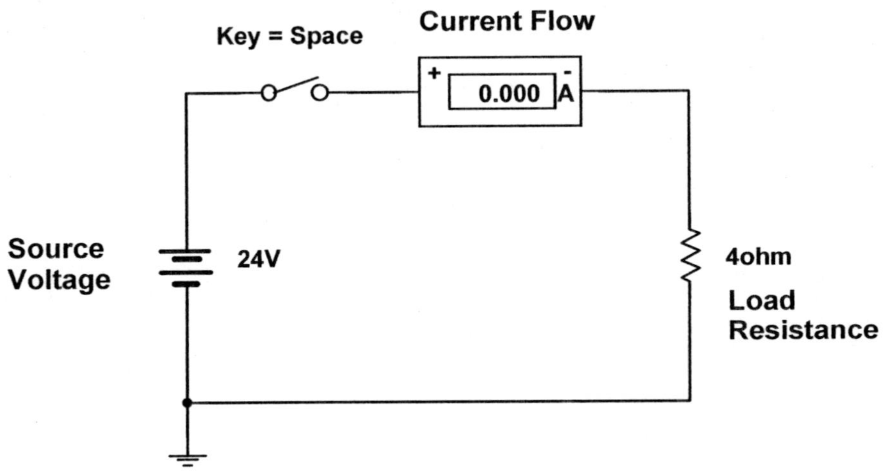

Answers

NAME ________________________________ DATE ________________

UNIT 6

ELECTRIC CONNECTIONS

Multisim Simulation Assignments

6-1 Open the Multisim circuit file CH06_01 shown.
One of the connectors (C1 or C2) has a higher than acceptable resistance, reducing the total energy normally available to the lamp.
(a) Turn the simulation "on", close switch S, and record the value of the voltage drop across connector C1, connector C2, and the lamp.
(b) Repeat with connector C1 bypassed with a jumper wire connected across it.
(c) Repeat with connector C2 bypassed with a jumper wire connected across it.
(d) Which connector has a higher than acceptable resistance? Why?

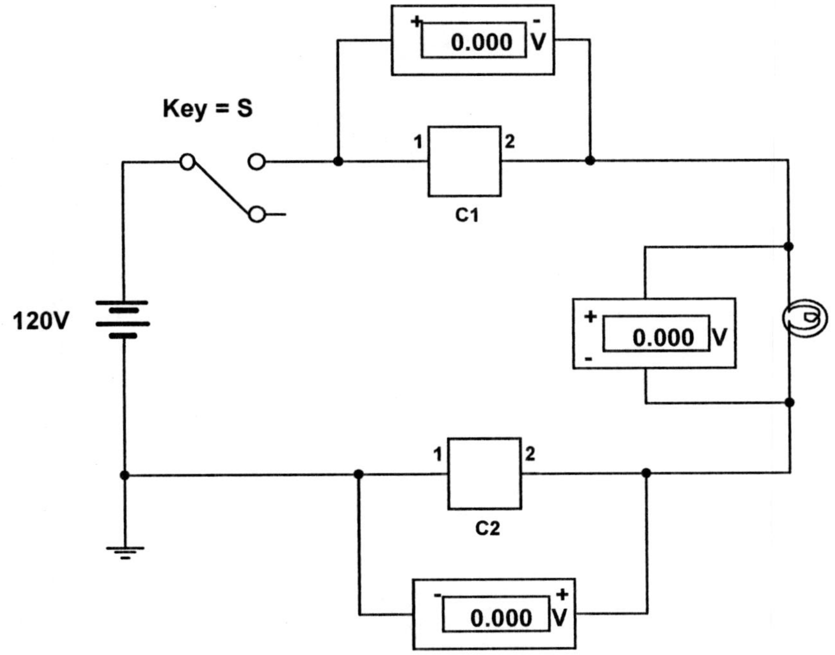

Answers

NAME ________________________________ DATE ________________

6-2 Open the Multisim circuit file CH06_02 shown.

(a) Complete the wiring from the switches through the cable to the lights, according to the wire number markings.

(b) Turn the simulation "on" and identify which lamp (X1, X2, X3, X4) is operated by which switch (A, B, C, D).

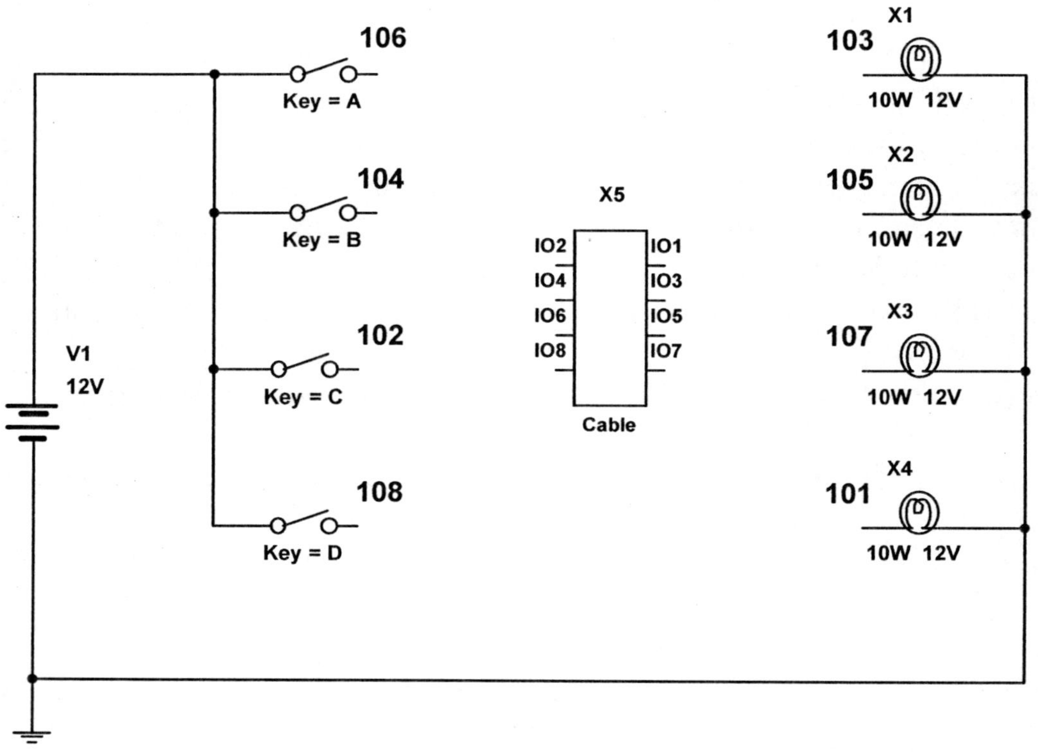

Answers

NAME ______________________________ DATE ________________

UNIT 7

SIMPLE, SERIES, AND PARALLEL CIRCUITS

Multisim Simulation Assignments

7-1 Open the Multisim circuit file CH07 _01 shown.
(a) Turn the simulation "on" and close switch "S".
Record the value of the current flow into and out of the lamp.
(b) Is the current flow into the lamp less than, greater than, or the same as the current flow out of the lamp?
(c) Record the value of the voltage across the lamp.
(d) Is the voltage across the lamp less than, greater than, or the same as the voltage of the source?

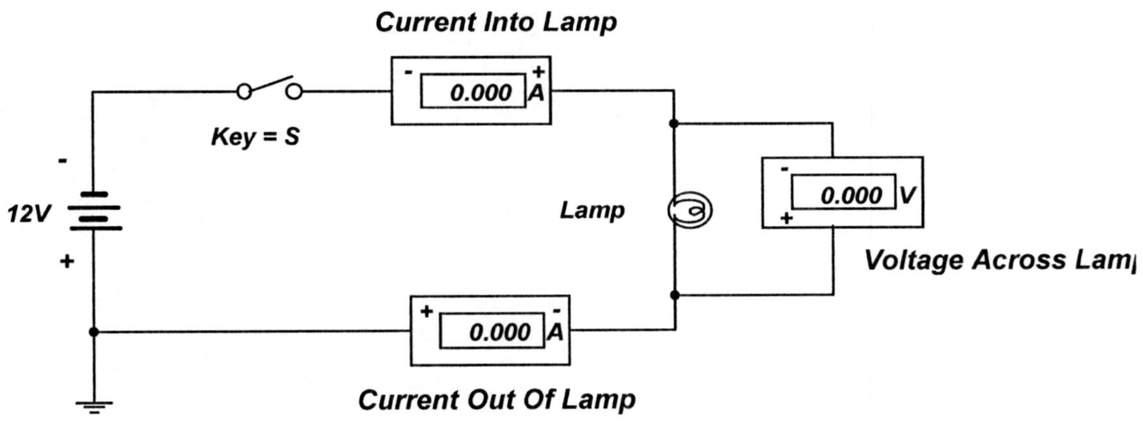

Answers

NAME ______________________________ DATE ______________

7-2 Open the Multisim circuit file CH07 _02 shown.

(a) What connection of lamps (series or parallel) is used?
Compare the ratings of the two lamps.

(b) Turn the simulation "on" and close switch "S".
Record the value of the voltage across each lamp.
How do these values compare with that of the source?

(c) Record the value of the current at the three points in the circuit.
How do these values compare with one another?

(d) Replace L2 with the "open lamp" and operate the circuit.
Does lamp L1 operate?
Compare the voltage across the open lamp with that of the source.

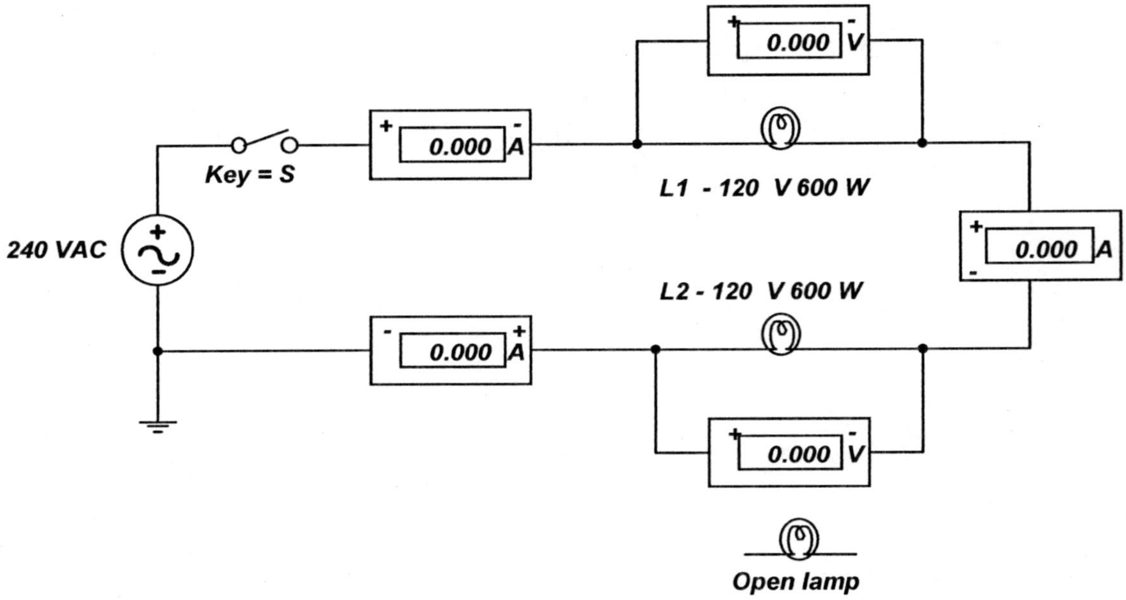

Answers

NAME ______________________________ DATE ________________

7-3 Open the Multisim circuit file CH07 _03 shown.

(a) What connection of lamps (series or parallel) is used?
Compare the ratings of the two lamps.

(b) Turn the simulation "on" and close switch "S".
Record the value of the voltage across each lamp.
How do these values compare with that of the source?

(c) Record the value of the line current, the current through L1 and the current through L2.
How do these values compare with one another?

(d) Replace L2 with the "open lamp" and operate the circuit.
Does lamp L1 operate normally?

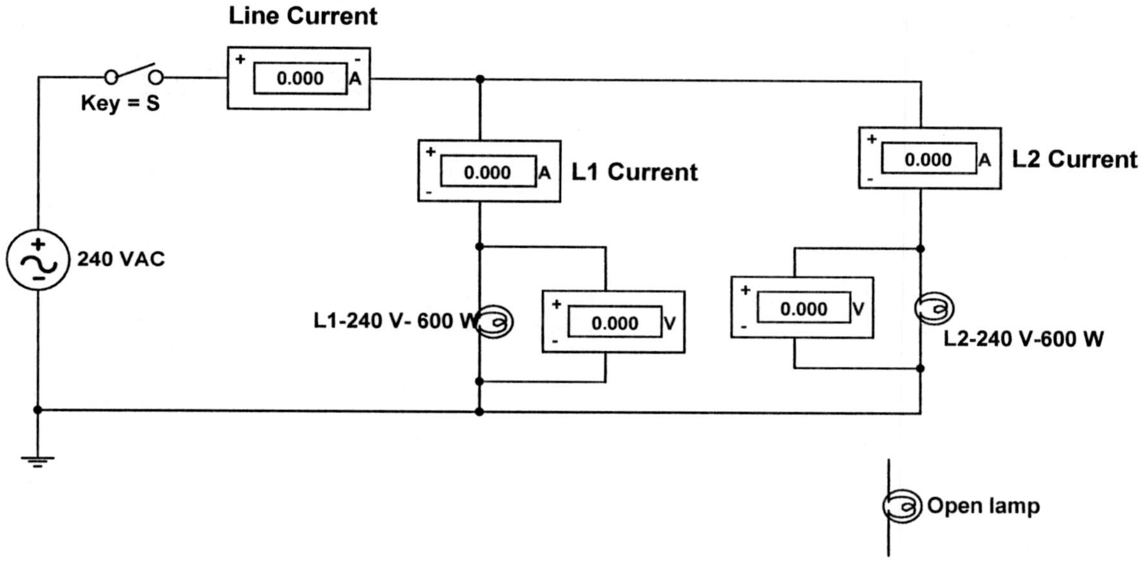

Answers

NAME ________________________________ DATE ________________

7-4 Open the Multisim circuit file CH07 _04 shown.

(a) The 12 V pilot light is to be operated from the 120 V source by connecting the resistor to lower the voltage applied to it. How should the resistor be connected relative to the lamp? Correctly connect the resistor and operate the circuit to verify your answer.

(b) What resistor connection would cause the pilot light to burn out? Incorrectly connect the resistor and operate the circuit to verify your answer.

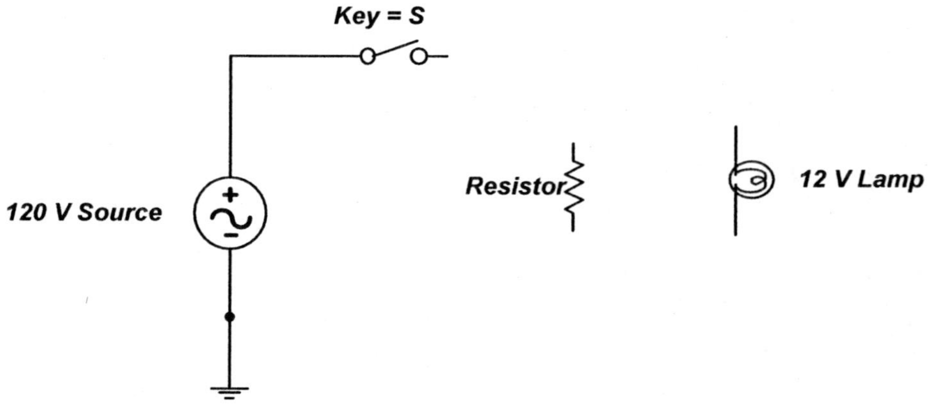

Answers

NAME ____________________________________ DATE __________________

7-5 Open the Multisim circuit file CH07 _05 shown.
A control system calls for a lamp to come on when switch and either switch B or switch C, is closed. What connection of switches will accomplish this? Correctly connect the switches and operate the circuit to verify your answer.

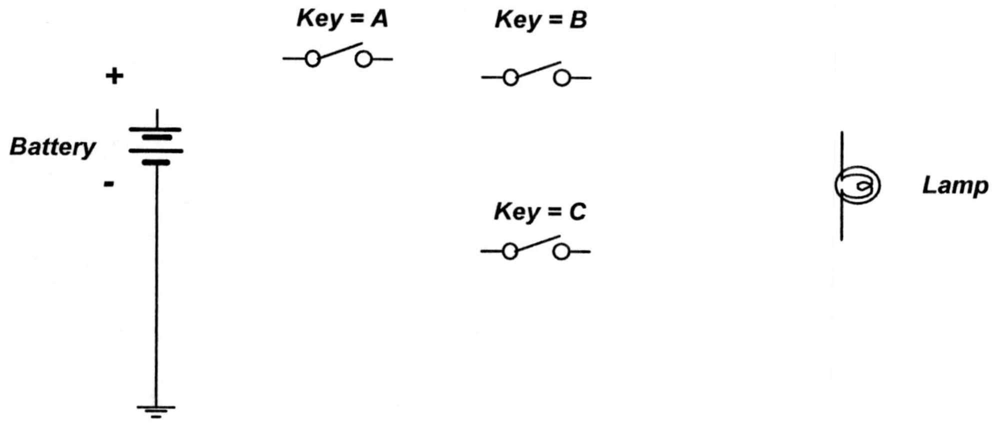

Answers

7-6 Open the Multisim circuit file CH07 _06 shown.
(a) What arrangement (series or parallel) of lamps is used?
(b) Use the spare lamp to determine which lamp is burnt open.

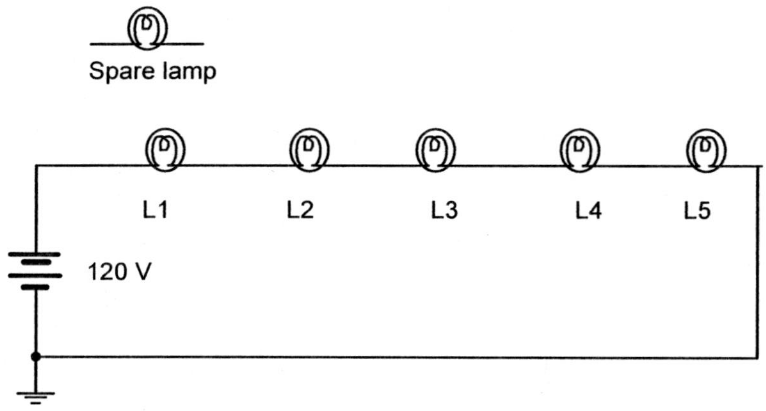

Answers

NAME ______________________________ DATE ________________

7-7 Open the Multisim circuit file CH07 _07 shown.

(a) Turn the simulation "on" and with switch "S" open record the value of the voltage across the switch and each lamp.

(b) With switch "S" closed record the value of the voltage across the switch and each lamp.

(c) Based on your readings, why must the three lamps be identical in resistance?

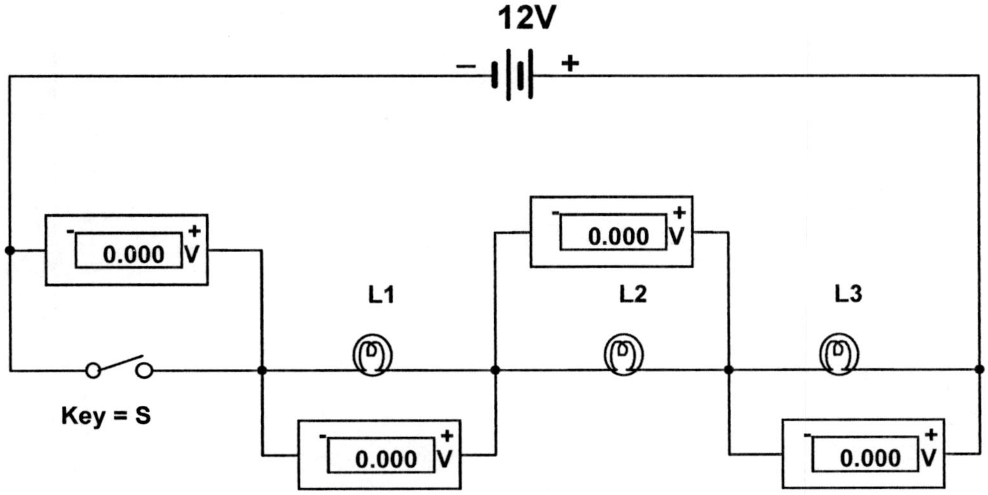

Answers

NAME ______________________________ DATE ______________

7-8 Open the Multisim circuit file CH07 _08 shown.
Turn the simulation "on" and with the switch closed observe the two voltmeter readings. Based on the voltmeter readings, which load (resistor or lamp) has the higher resistance? Why?

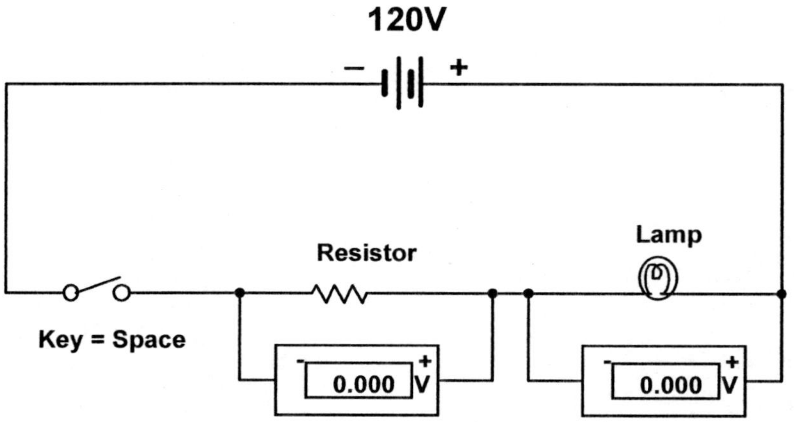

Answers

NAME ______________________________________ DATE ____________________

7-9 Open the Multisim circuit file CH07 _09 shown.
Turn the simulation "on" and operate the circuit. According to the voltmeter readings what is the value of the source voltage? Why?

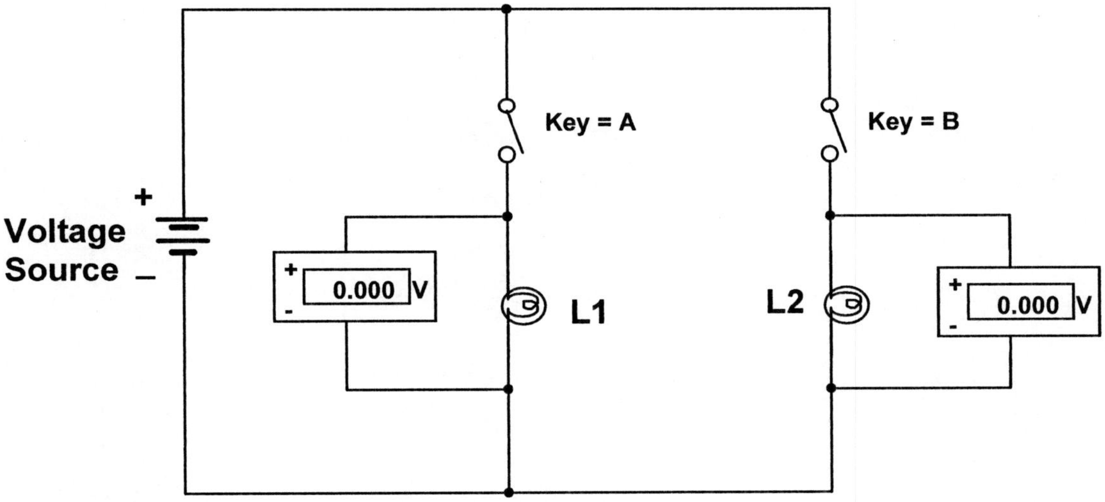

Answers

NAME ______________________________ DATE ________________

7-10 Open the Multisim circuit file CH07 _10 shown.
Turn the simulation "on" and operate the circuit.
(a) Record the amount of current drawn by the 100 W lamp.
(b) Record the amount of current drawn by the 40 W lamp.
(c) Which lamp has the highest resistance? Why?

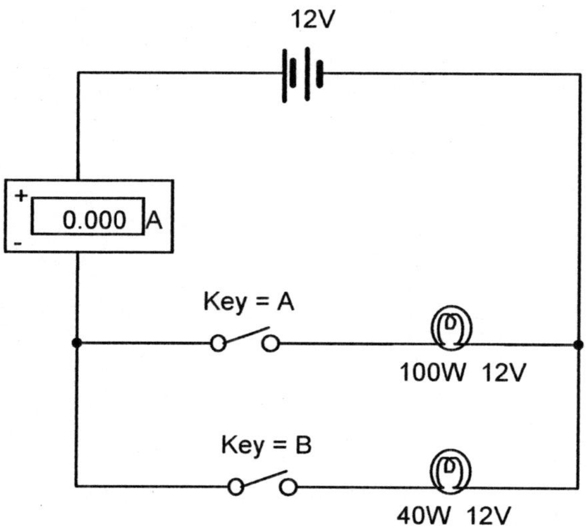

Answers

NAME ____________________ DATE ____________

7-11 Open the Multisim circuit file CH07 _11 shown.
Complete the wiring for the three lamps connected in series with the battery. Turn the simulation "on" and operate the circuit. Simulate the removal of lamp 3 from its socket by disconnecting its terminal leads. What effect does this have on the other lamps? Why?

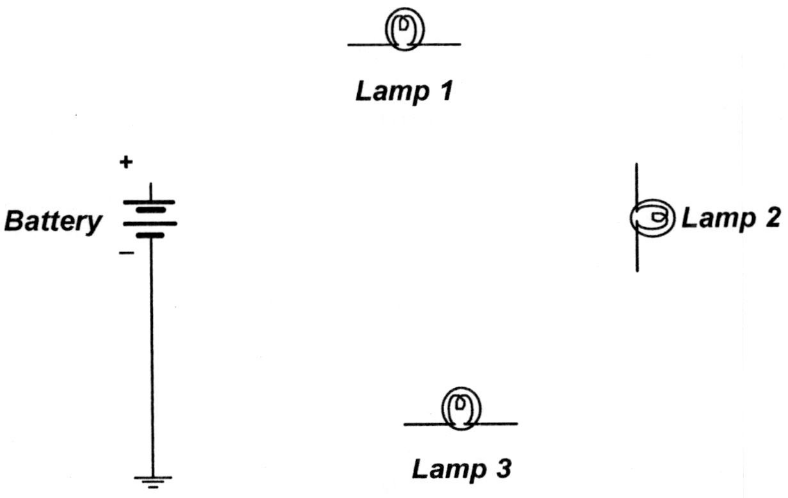

Answers

NAME ____________________ DATE ______________

7-12 Open the Multisim circuit file CH07 _12 shown.
Complete the wiring for the three lamps connected in parallel with the battery. Turn the simulation "on" and operate the circuit. Simulate the removal of lamp 2 from its socket by disconnecting its terminal leads. What effect does this have on the other lamps? Why?

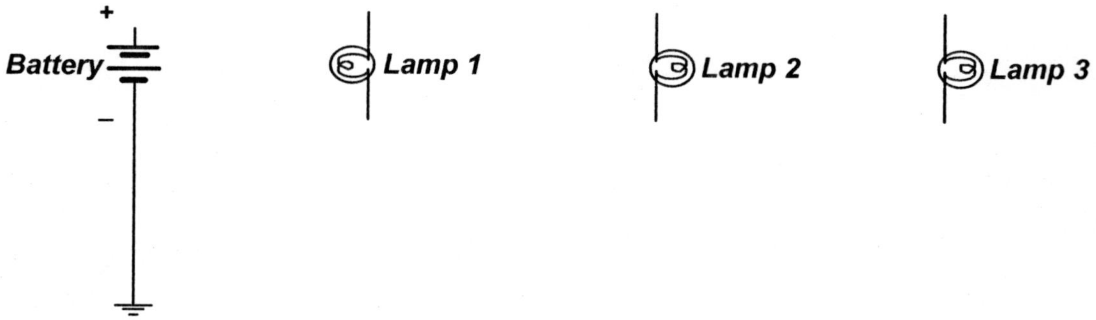

Answers

NAME ________________________ DATE ______________

7-13 Open the Multisim circuit file CH07 _13 shown.
Complete the wiring for switches (A and B) in series to control the lamp. Turn the simulation "on" and operate the circuit. Construct a truth table that shows how the switches control the lamp.

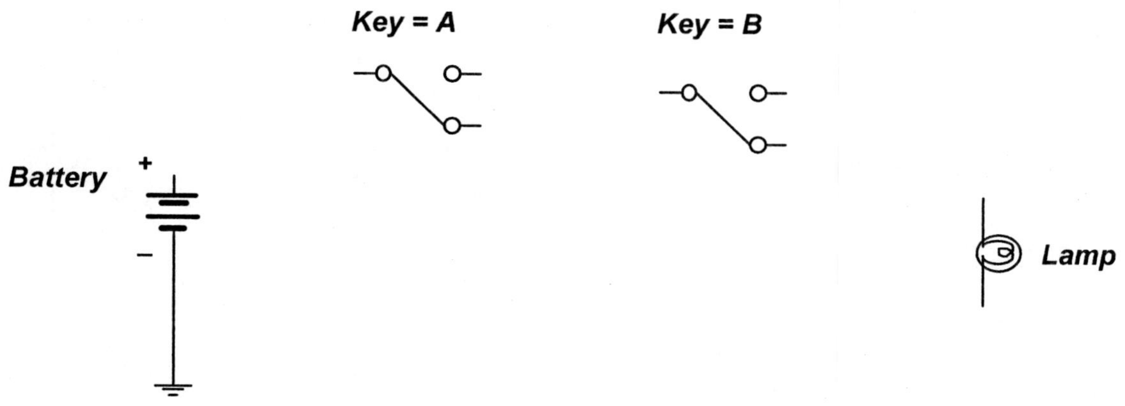

Answers

NAME ________________________________ DATE ____________________

7-14 Open the Multisim circuit file CH07 _14 shown.
Complete the wiring for switches (A and B) in parallel to control the lamp. Turn the simulation "on" and operate the circuit. Construct a truth table that shows how the switches control the lamp.

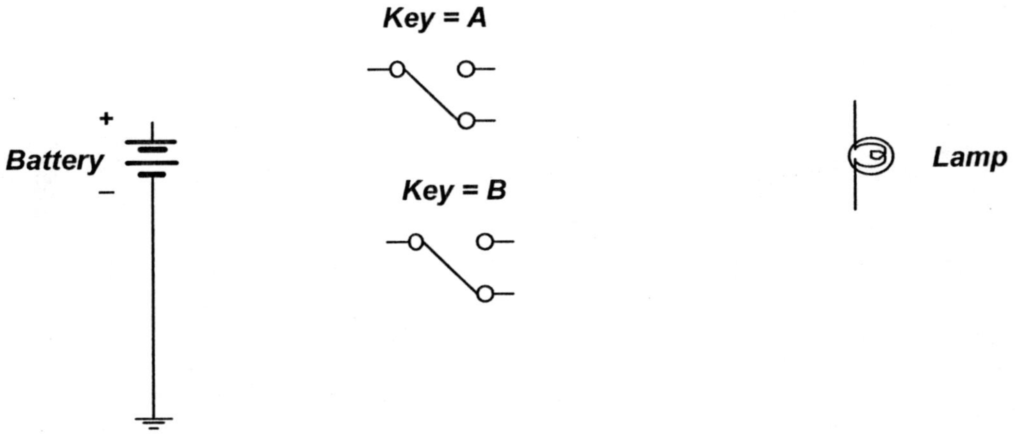

Answers

NAME ______________________________ DATE ________________

7-15 Open the Multisim circuit file CH07 _15 shown.
Turn the simulation "on" and operate the circuit. Construct a truth table that shows how the switches control the lamp.

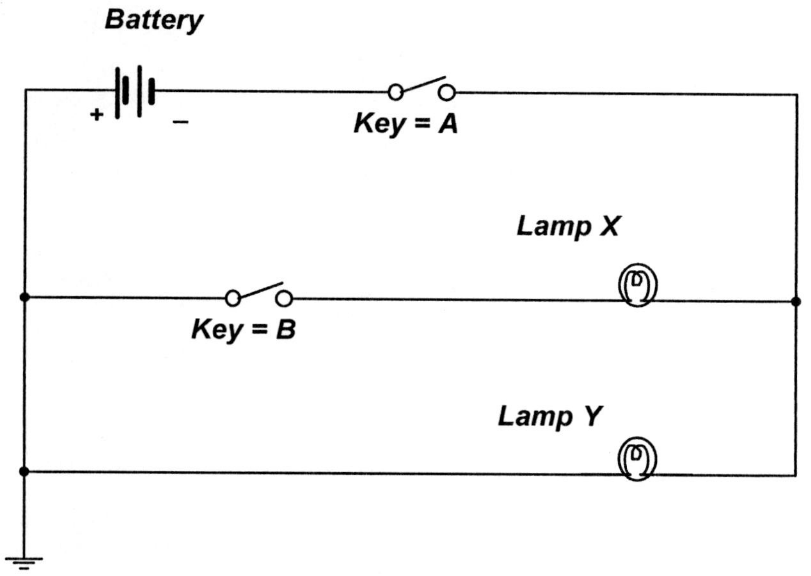

Answers

NAME ______________________________ DATE __________________

7-16 Open the Multisim circuit file CH07 _16 shown.
Turn the simulation "on" and operate the circuit.

(a) In what way is the wiring of S1 to control L1 different from that used for S2 to control L2?

(b) What difference, if any, does this have on the way each switch controls each light. Why?

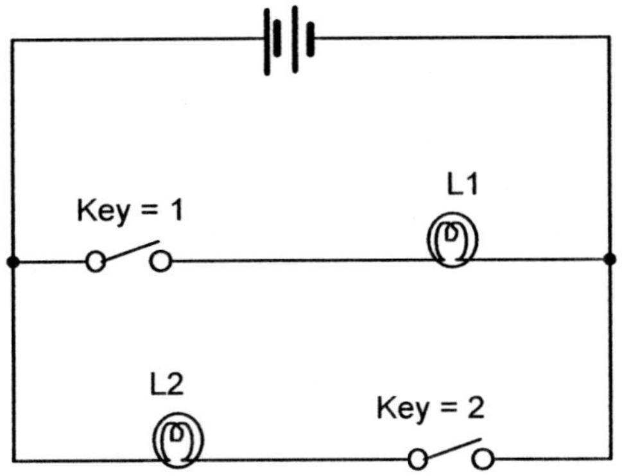

Answers

NAME ______________________________ DATE ________________

UNIT 8

MEASURING VOLTAGE, CURRENT, AND RESISTANCE

Multisim Simulation Assignments

8-1 Open the Multisim circuit file CH08 _01 shown.
Assume that a DC digital voltmeter is connected so that the positive lead of the meter connects to the negative side of the circuit. What will happen? Verify your answer by reconnecting the digital voltmeter to simulate this condition.

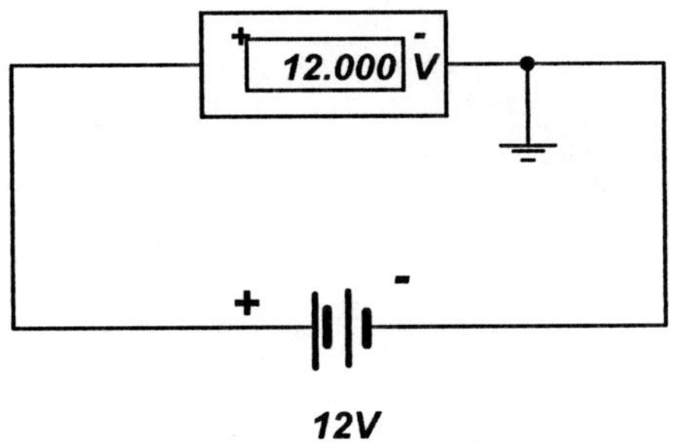

Answers

NAME ____ DATE ____________________

8-2 Open the Multisim circuit file CH08 _02 shown.

An ohmmeter is connected across the two leads of a single pole toggle switch to check its continuity out of circuit. If the switch is turned on and off, what ohmmeter reading should be indicated for each of the following conditions: switch in "on" and switch in "off" position?

(a) Switch operation normal

(b) Switch is defective open

(c) Switch is defected shorted

Verify your answers by turning the simulation "on" to test each switch using the ohmmeter.

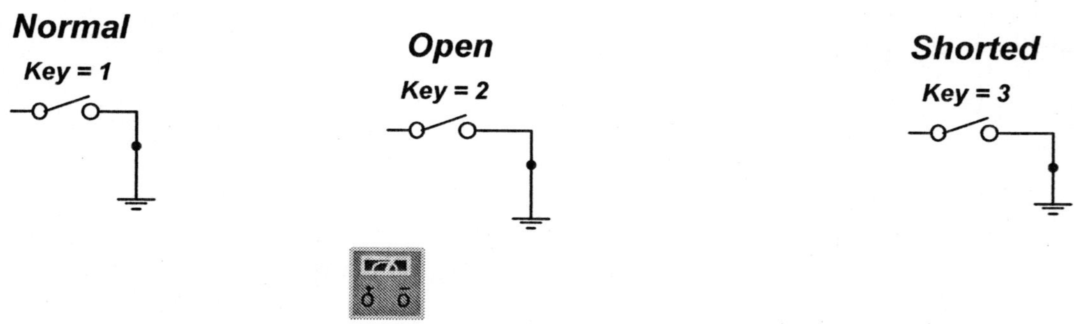

Answers

NAME ______________________________ DATE ________________

8-3 Open the Multisim circuit file CH08 _03 shown.

(a) Assume that while measuring current in the circuit shown, the two test leads of the ammeter are accidentally shorted together. What effect (if any)would this have on the circuit under test? Why? Verify your answer by turning the simulation "on" and creating a short across the ammeter.

(a) Assume that while measuring voltage in the circuit shown, the two test leads of the voltmeter are accidentally shorted together. What effect (if any)would this have on the circuit under test? Why? Verify your answer by turning the simulation "on" and creating a short across the voltmeter.

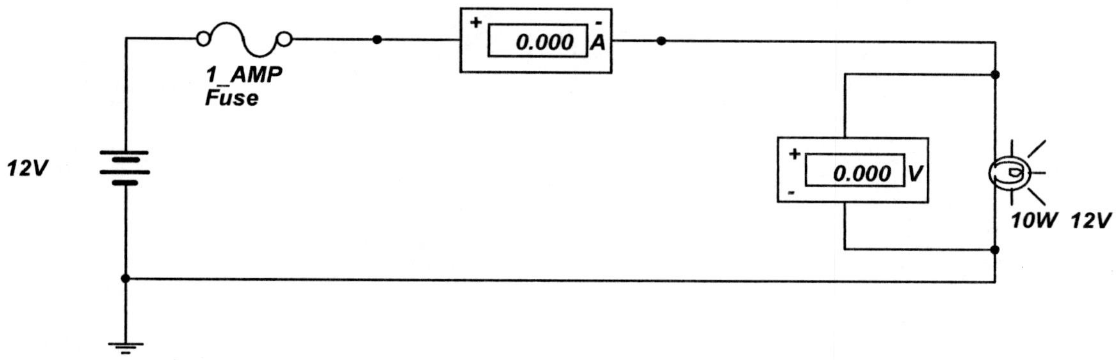

Answers

NAME ___________________________________ DATE _________________

8-4 Open the Multisim circuit file CH08 _04 shown.
Turn the simulation "on" and operate the circuit. Use the multimeter to measure the voltage between the test points A, B, C and the common ground:
(a) With the switch "off".
(b) With the switch "on".

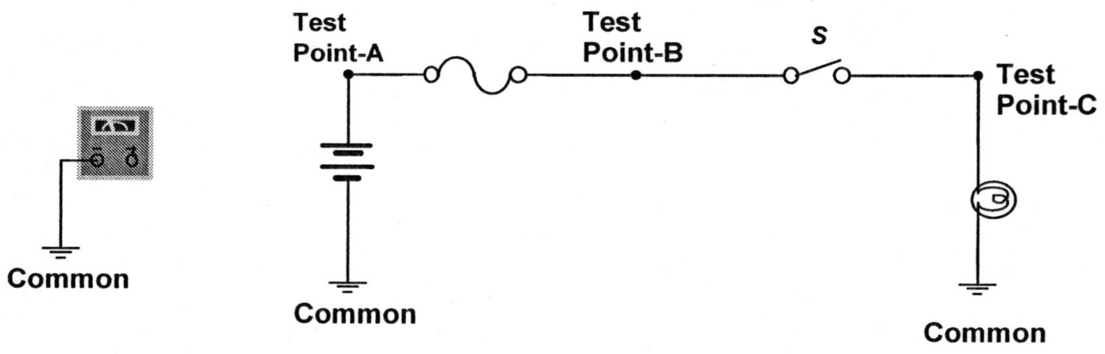

<u>Answers</u>

NAME ____ DATE ____________________

8-5 Open the Multisim circuit file CH08 _05 shown.

Load R1 is suspected of having a lower than normal resistance value. Turn the simulation "on" and operate the circuit. Use the multimeter to measure the actual resistance value of load R1.

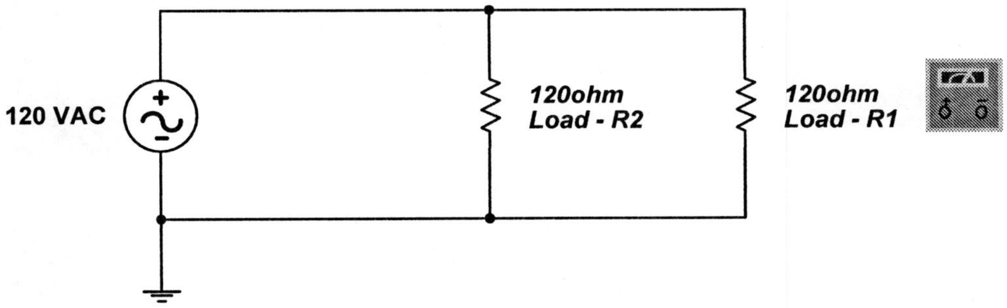

Answers

NAME ________________________________ DATE ________________

8-6 Open the Multisim circuit file CH08 _06 shown.
Turn the simulation "on" and operate the circuit.
(a) Use the multimeter to measure the value of the voltage applied to the electronic control module.
(b) Connect the + test lead of the meter to ground. What happens? Why?

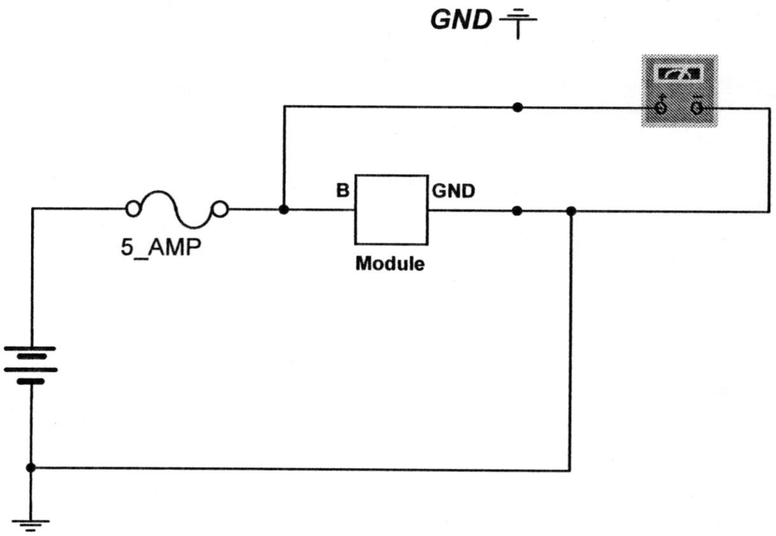

Answers

NAME ______________________________ DATE ______________

8-7 Open the Multisim circuit file CH08 _07 shown.
Turn the simulation "on" and operate the circuit. Connect the multimeter and record the value and type of voltage (AC or DC) obtained from each of the power supply sources shown:

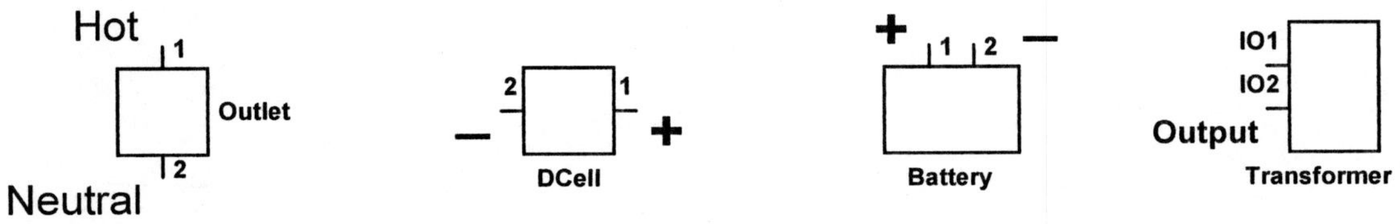

(a) (b) (c) (d)

Answers

NAME ______________________________ DATE ______________

8-8 Open the Multisim circuit file CH08 _08 shown.
Turn the simulation "on" and operate the circuit. Connect the multimeter and determine:
(a) The polarity of terminals T1 and T2 of DC voltage source A.
(b) The polarity of terminals T1 and T2 of DC voltage source B.

DC Voltage Source A

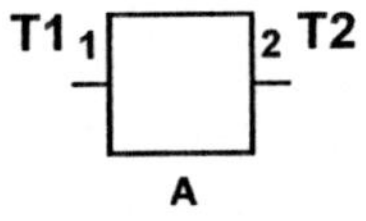

DC Voltage Source B

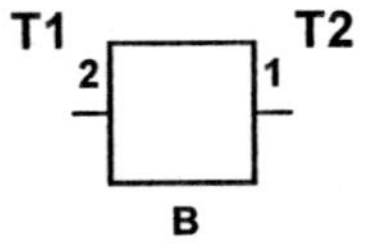

Answers

NAME ______________________________________ DATE ____________________

8-9 Open the Multisim circuit file CH08 _09 shown.
Turn the simulation "on" and operate the series circuit shown.
(a) With the switch closed use the multimeter to measure the:
- (i) battery voltage
- (ii) voltage drop across the fuse
- (iii) voltage drop across the switch
- (iv) voltage drop across lamp L1
- (v) voltage drop across lamp L2

(b) Which lamp (L1 or L2) has the greater resistance? Why?

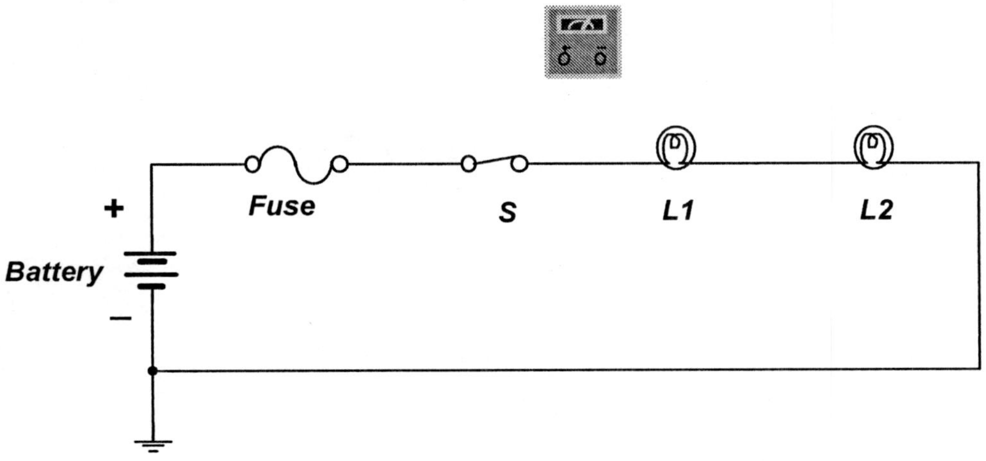

Answers

NAME ______________________________ DATE ________________

8-10 Open the Multisim circuit file CH08 _10 shown.
Use the multimeter to record the amount of current that flows when 240 VAC is applied in turn to each of the following loads:
(a) 4 ohm
(b) 10 ohm
(c) 24 ohm
(d) 100 ohm

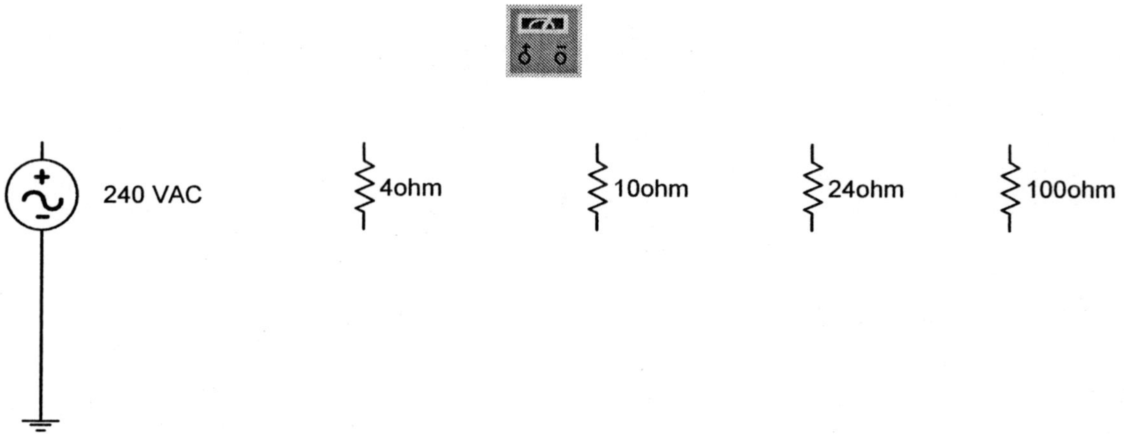

Answers

NAME ______________________________ DATE ________________

8-11 Open the Multisim circuit file CH08 _11 shown.
Use the multimeter to record the resistance of each of the following loads: (a) R1
(b) R2
(c) R3
(d) R4
(e) R5

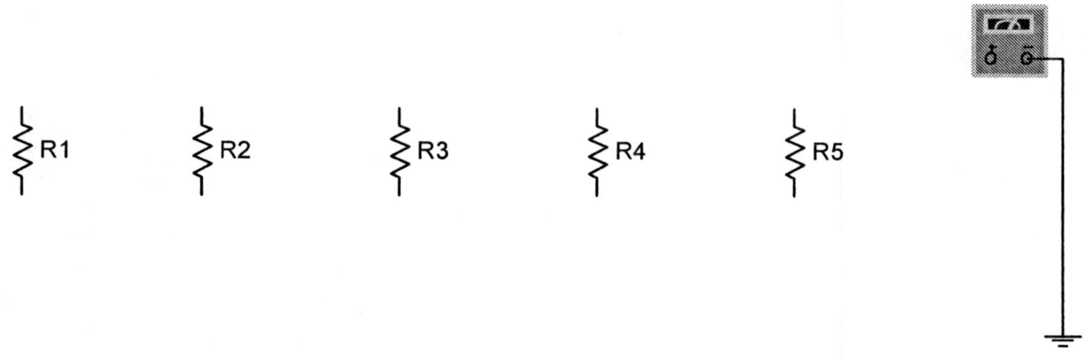

Answers

NAME ______________________________ DATE ________________

8-12 Open the Multisim circuit file CH08 _12 shown.
Use the multimeter to perform a continuity test of each component and classify each as being OK or defective.

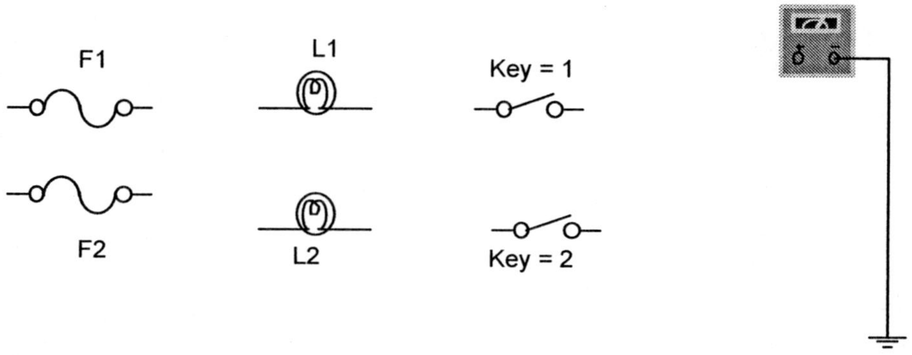

Answers

NAME ______________________________ DATE ______________

8-13 Open the Multisim circuit file CH08 _13 shown.
Turn the simulation "on" and operate the parallel circuit shown.
Record the total current for each of the following conditions:
(a) S1 closed – S2 & S3 open
(b) S2 closed – S1 & S3 open
(c) S3 closed – S1 & S2 open
(d) S1 & S2 closed – S3 open
(e) S1 & S3 closed – S2 open
(f) S3 & S2 closed – S1 open
(g) S1, S2, and S3 closed

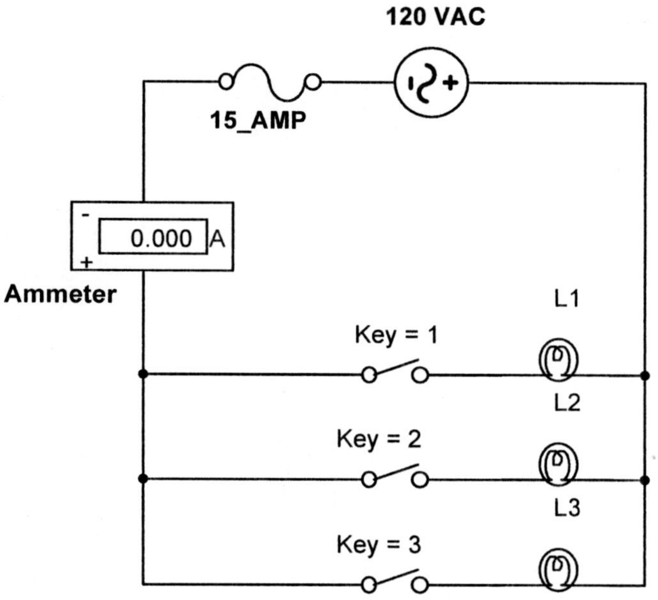

Answers

NAME ______________________________ DATE ________________

8-14 Open the Multisim circuit file CH08 _14 shown.
(a) What connection of lamps (series or parallel) is used?
(b) Connect the ammeter and record the amount of current flow.
(c) Connect the voltmeter and record the amount of voltage drop across each lamp.

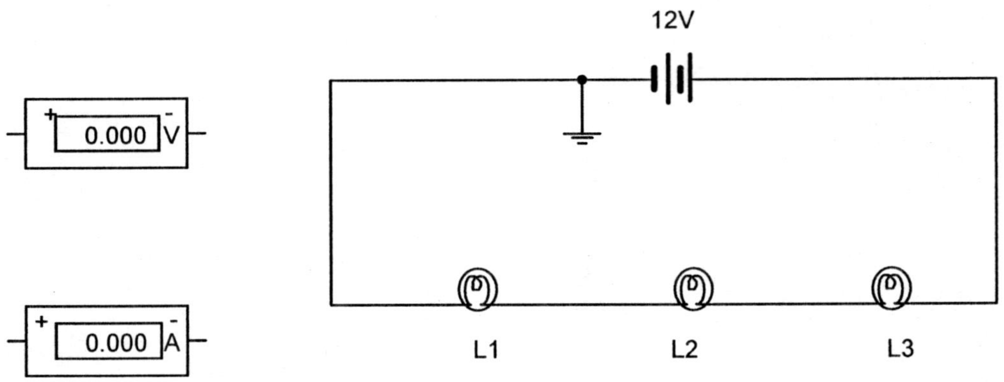

Answers

NAME ___________________________ DATE _______________

8-15 Open the Multisim circuit file CH08 _15 shown.
Turn the simulation "on" and operate the circuit.
Connect the multimeter to measure:
(a) The DC source voltage
(b) The voltage drop across R1, L1, and L2.
(c) The current flow through R1, L1, and L2.

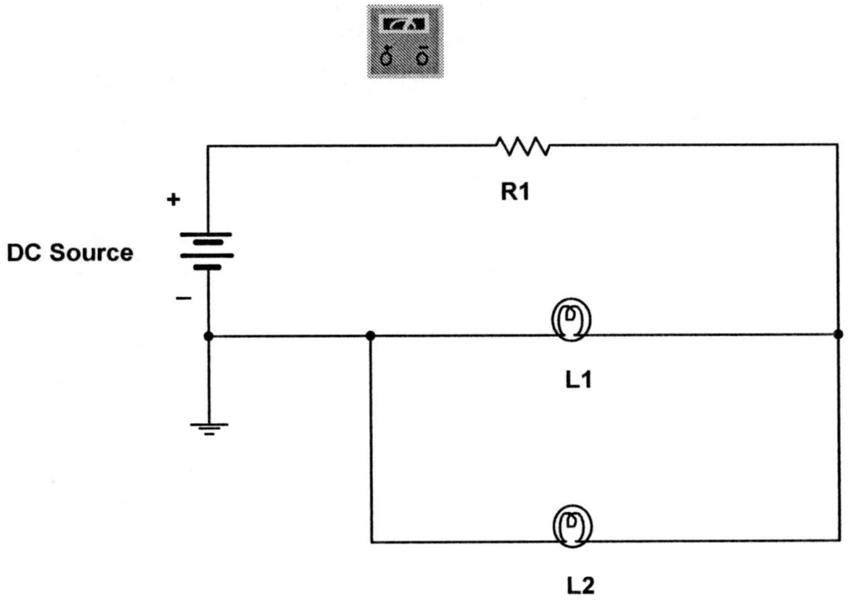

Answers

NAME ______________________________ DATE ________________

UNIT 9

OHM'S LAW

Multisim Simulation Assignments

9-1 Open the Multisim circuit file CH09 _01 shown.
- (i) For each circuit drawn [(a) through (e)], calculate the expected current flow according to Ohm's law
- (ii) Use the multimeter to measure the actual amount of current flow for each circuit.

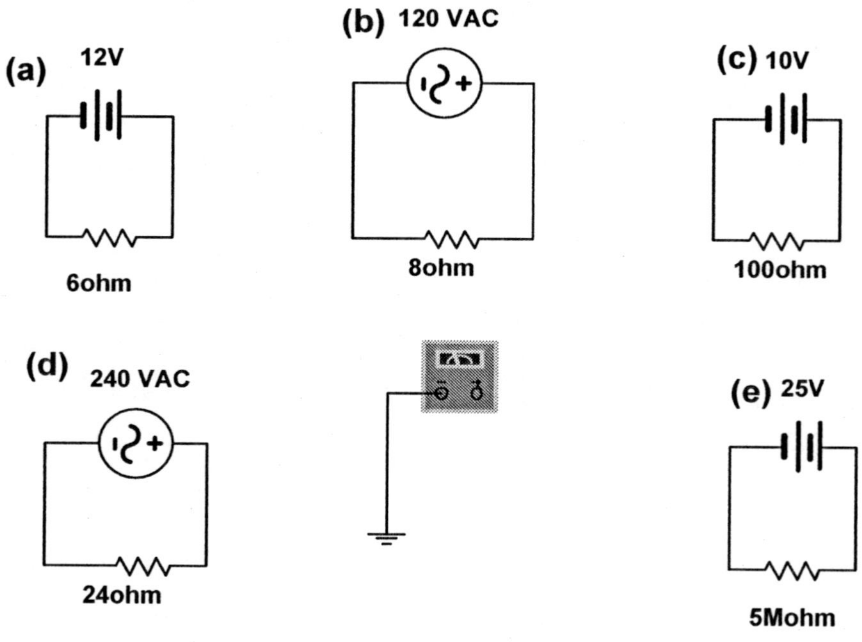

<u>Answers</u>

NAME ______________________________ DATE ________________

9-2 Open the Multisim circuit file CH09 _02 shown.

(a) Connect each resistor, in turn, to the 100-volt source. Turn the simulation "on" and record the amount of circuit current flow for each resistance value.

(b) According to your results, explain the effect the changes in circuit resistance have on the current flow.

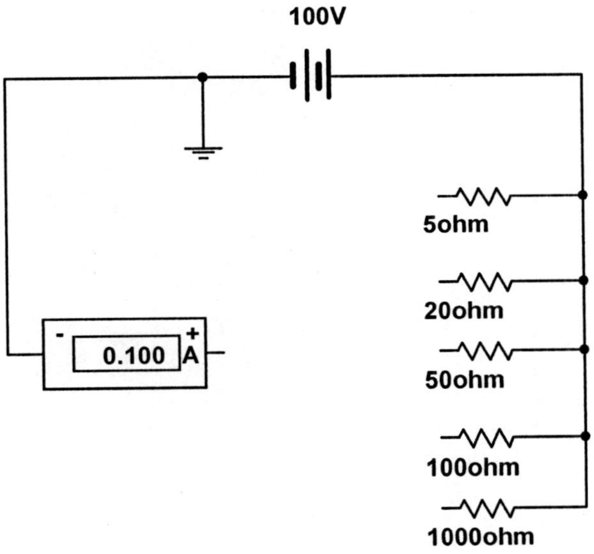

Answers

NAME ____________________________________ DATE ____________________

9-3 Open the Multisim circuit file CH09 _03 shown.

(i) For each circuit shown, calculate the expected applied voltage according to Ohm's law.

(ii) Use the multimeter to measure the actual amount of voltage applied to each circuit.

(a)

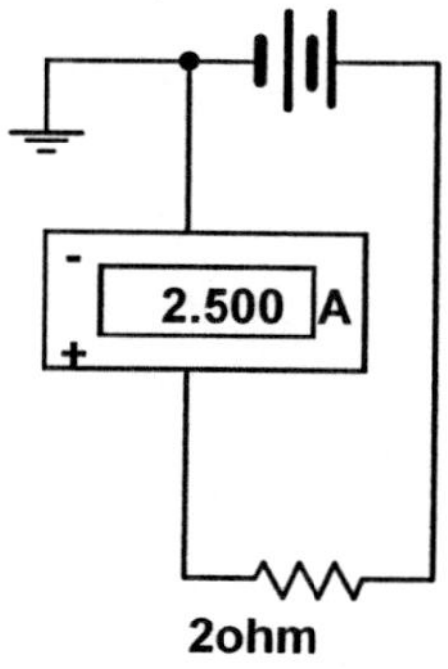

(b)

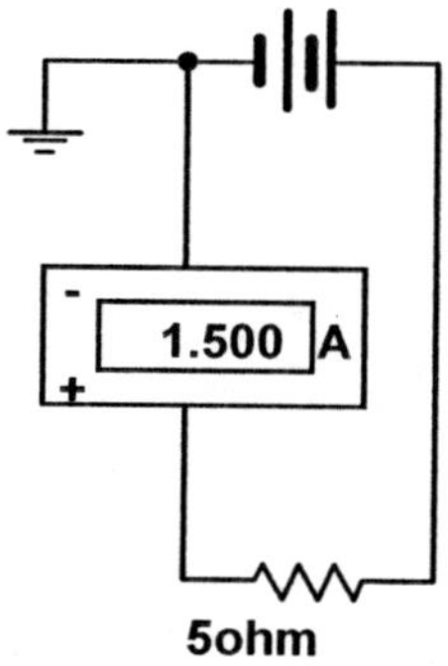

(c)

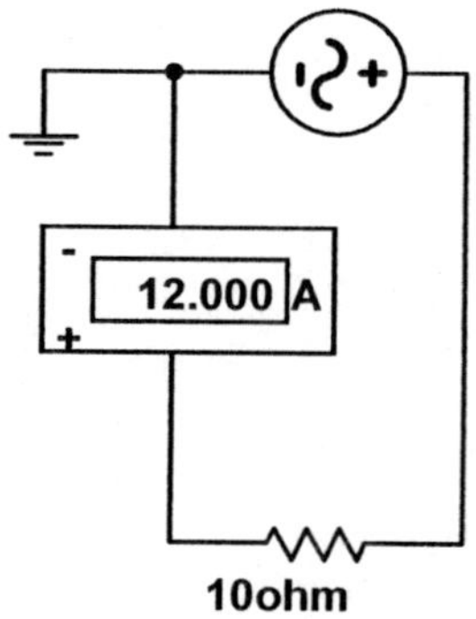

(d)

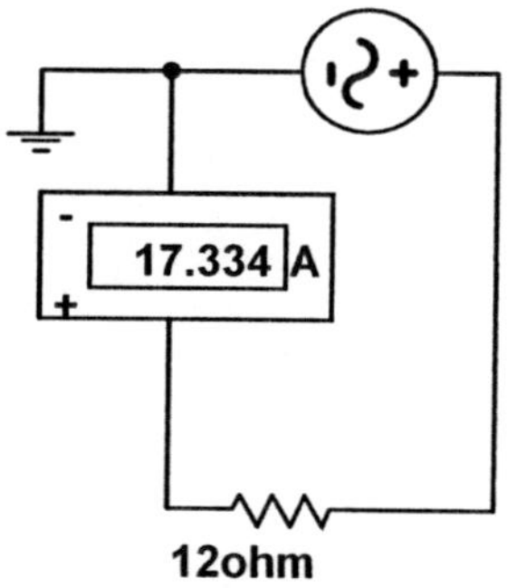

(e)

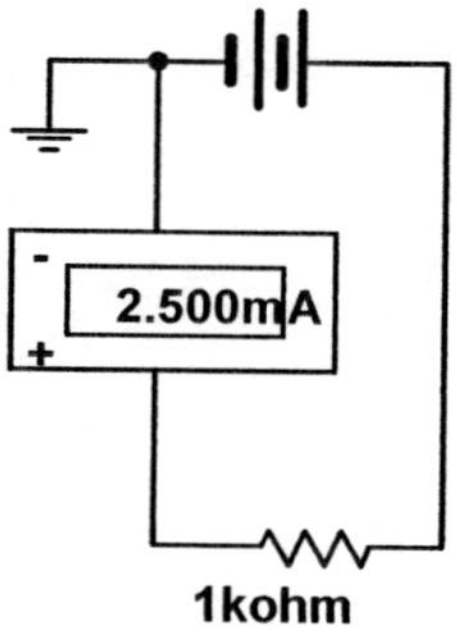

Answers

NAME ______________________________ DATE ____________________

9-4 Open the Multisim circuit file CH09 _04 shown.

(i) For each circuit shown, calculate the expected load resistance according to Ohm's law.

(ii) Use the multimeter to measure the actual resistance of the load.

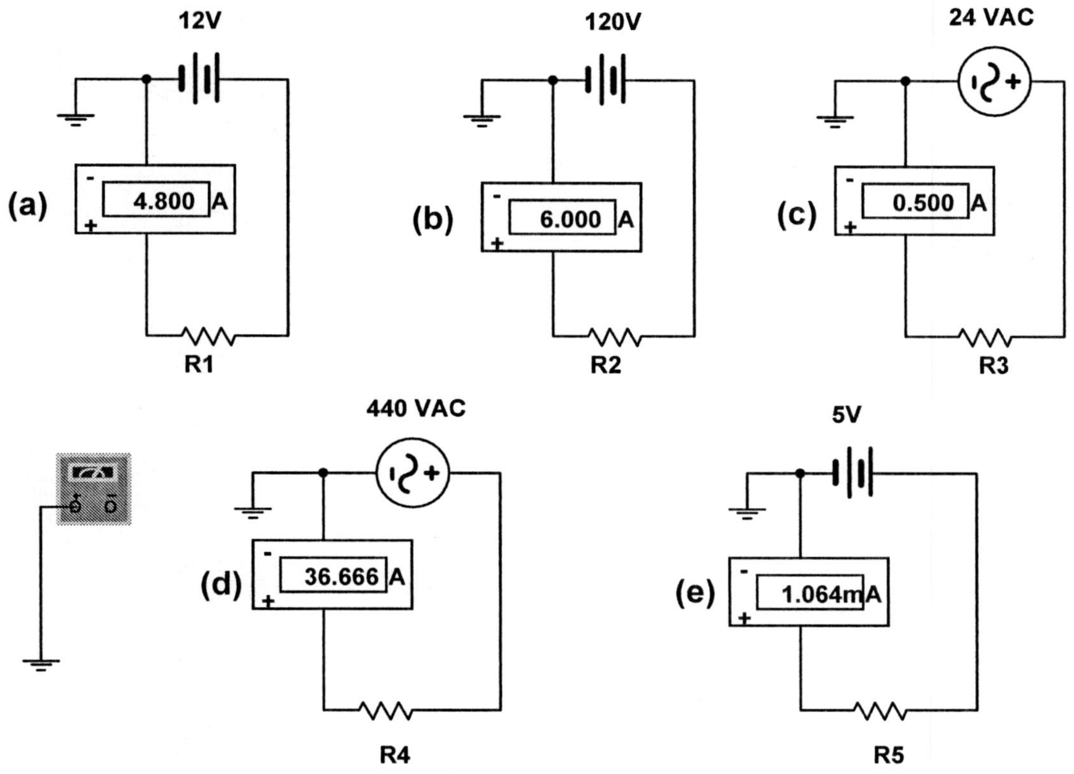

Answers

NAME ______________________________ DATE ________________

9-5 Open the Multisim circuit file CH09 _05 shown.

(a) Change the supply voltage (in 1-V steps) from 0 to 10 volts. Record the amount of current for each voltage step.

(b) According to your results, explain the effect on current of changes in voltage.

(c) Plot a graph of current versus voltage. With reference to this graph, explain why the resistor would be classified as a linear device.

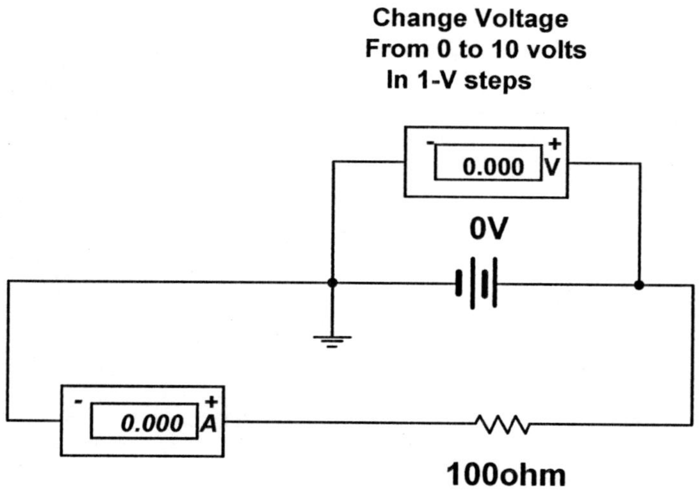

Answers

NAME ______________________________ DATE ________________

9-6 Open the Multisim circuit file CH09 _06 shown.
Turn the simulation "on" and observe the meter readings for each of the three circuits. Using the different power formulas compute the:
(a) wattage of the lamp
(b) power dissipated by the resistor
(c) power rating of the portable heater

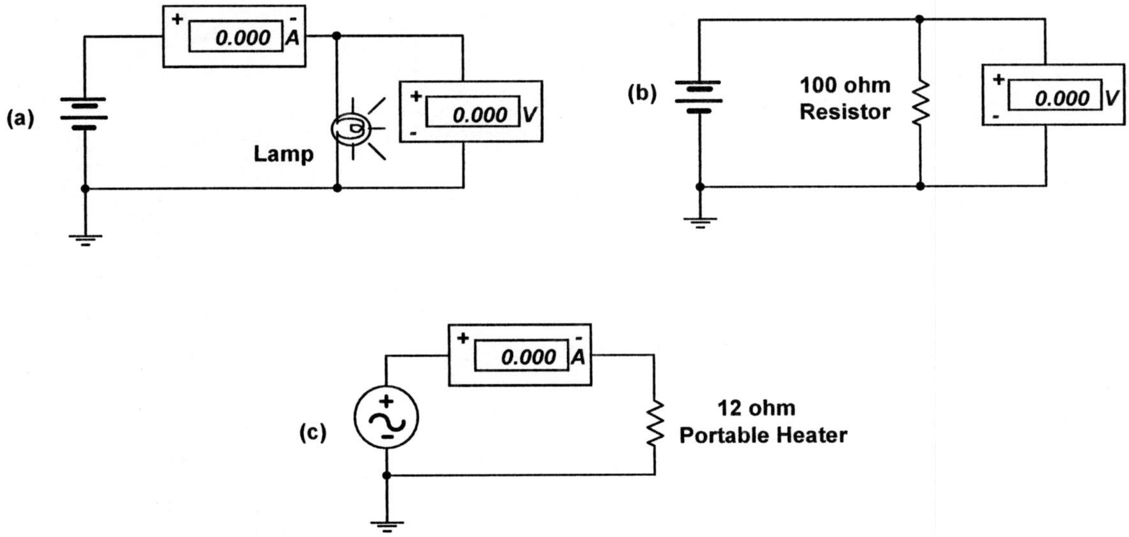

<u>Answers</u>

NAME ______________________ DATE ______________

9-7 Open the Multisim circuit file CH09 _07 shown.
Turn the simulation "on" and observe the ammeter reading.
(a) Calculate the value of the voltage drop across the resistor.
(b) Use the multimeter to measure the actual value of the voltage drop across the resistor.

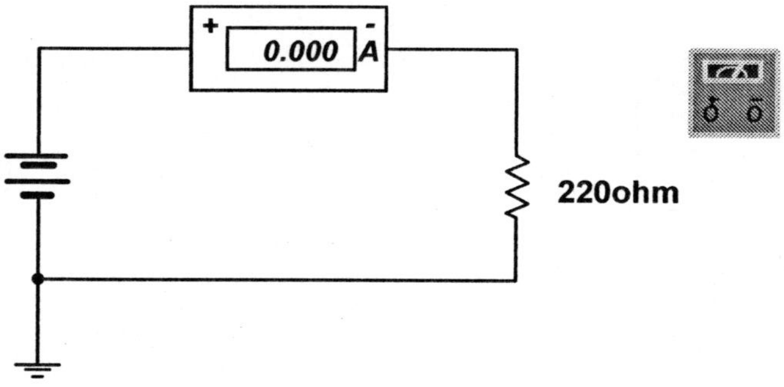

Answers

NAME ______________________________ DATE ________________

9-8 Open the Multisim circuit file CH09 _08 shown.

(a) Calculate the value of the current flow through module.

(b) Use the multimeter to measure the actual value of the current flow through the module.

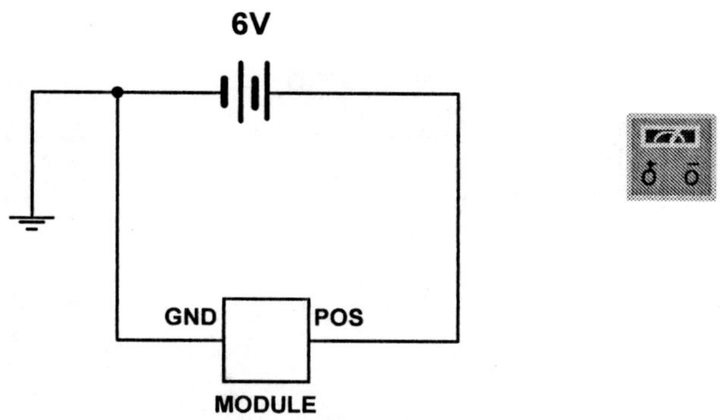

Resistance = 50,000 ohms

Answers

NAME ______________________________ DATE ____________________

9-9 Open the Multisim circuit file CH09 _09 shown.
Turn the simulation "on" and observe the meter readings.
(a) Calculate the value of resistor R.
(b) Use the multimeter to measure the actual value of resistor R.

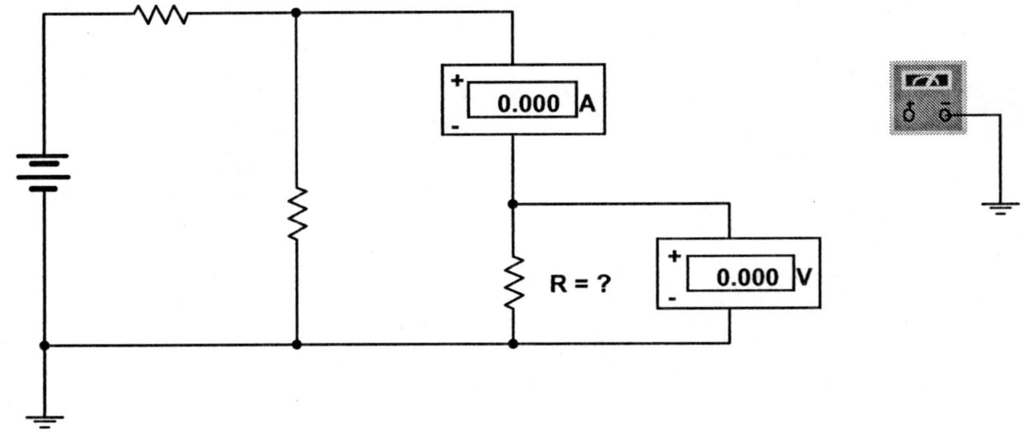

Answers

NAME ______________________________ DATE ________________

9-10 Open the Multisim circuit file CH09 _10 shown.

(a) Turn the simulation "on" and record the value of the circuit voltage, current, and power.

(b) What is the relationship between the volts, amperes, and watts in this circuit?

(c) Decrease the value of the load resistance from 20 Ω to 10 Ω. What effect does this have on the voltage, current, and power?

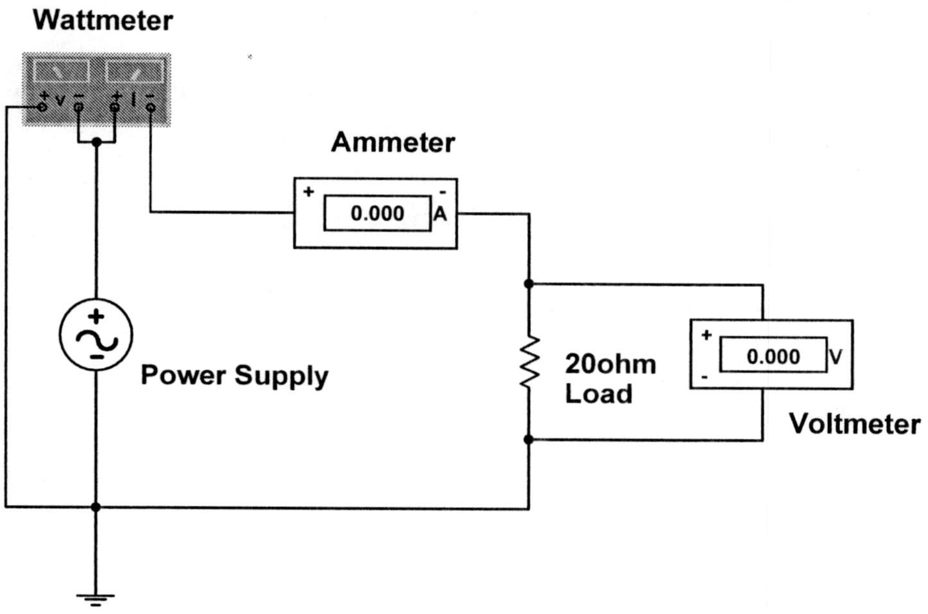

Answers

9-11 Open the Multisim circuit file CH09 _11 shown.
Ohm's law can be used to diagnose why circuits are not working as they should be. This lab exercise is designed to show how Ohm's law can be used to troubleshoot a circuit. Of the three circuits shown only one is operating normally.

(a) Compute the value of the normal current flow for each circuit according to the values specified.

(b) Use the multimeter to measure the actual current flow in each circuit.

(c) Based on your readings, identify which circuit is operating normally.

(d) Which circuit is operating at lower than normal current? Use the multimeter to determine why.

(e) Which circuit is operating at higher than normal current? Use the multimeter to determine why.

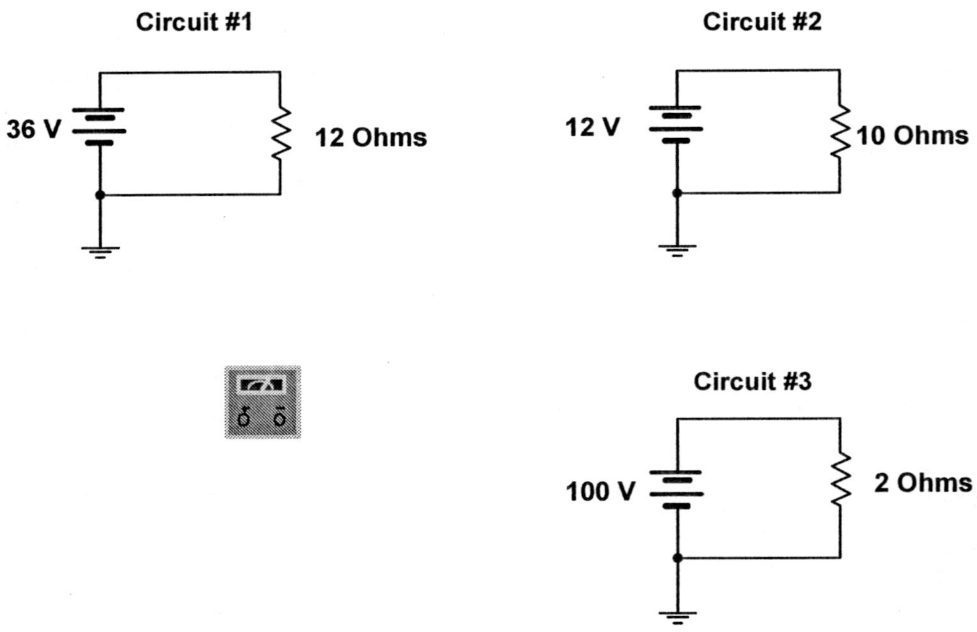

Answers

NAME ____________________ DATE ____________

UNIT 10

CIRCUIT CONDUCTORS AND WIRE SIZES

Multisim Simulation Assignments

10-1 Open the Multisim circuit file CH10 _01 shown.
- (a) Turn the simulation "on" and record the value of the source and load voltages with the load switched "off".
- (b) Repeat part (a) with the load switched "on".
- (c) Determine the value of the line voltage drop.
- (d) Calculate the percent voltage drop.
- (e) According to Ohm's Law why is the voltage drop zero when the load is not switched "on"?

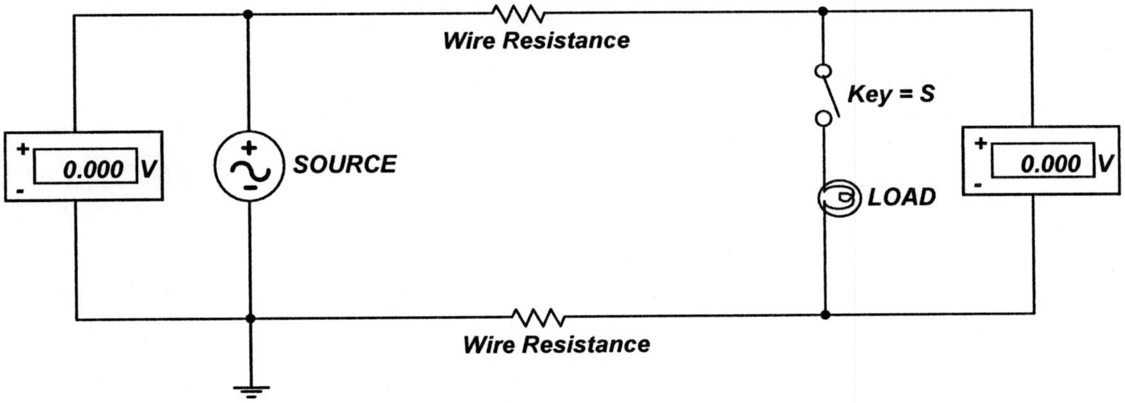

Answers

NAME ________________________ DATE ________________

10-2 Open the Multisim circuit file CH10 _02 shown.

(a) Turn the simulation "on" and record the value of the load current.
(b) Use Ohm's Law to calculate the expected line voltage drop.
(c) Use the multimeter to measure the operating voltage across the load.
(d) Calculate the actual line voltage drop.

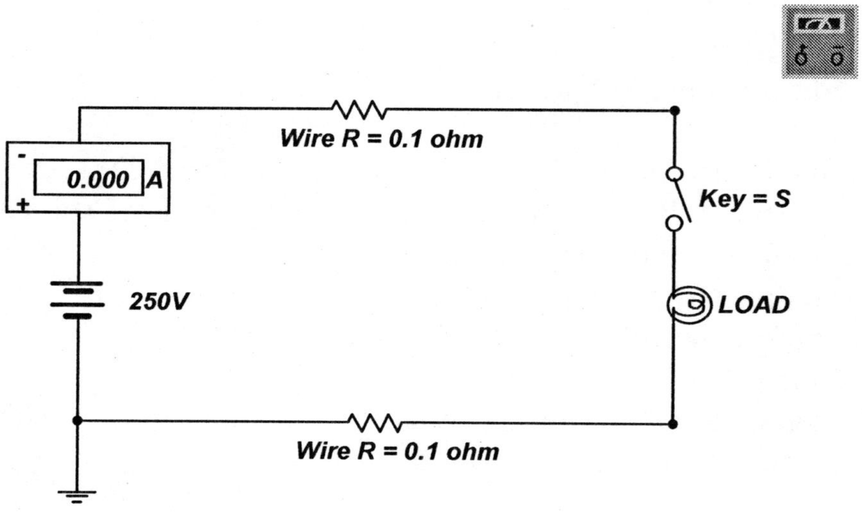

Answers

NAME ______________________________________ DATE __________________

10-3 Open the Multisim circuit file CH10 _03 shown.

(a) Calculate the resistance of the 1,000 ft of AWG 10 copper using a resistivity K of 12.9 ohms per circular mil foot.

(b) Turn the simulation "on" and use the multimeter to record the value of the length of 1,000 ft of AWG 10 copper wire.

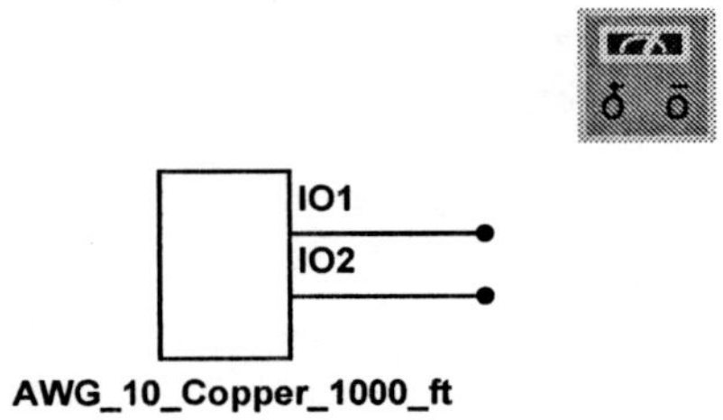

Answers

NAME ____ DATE ____________________

10-4 Open the Multisim circuit file CH10 _04 shown.

(a) Turn the simulation "on" and use the multimeter to measure the resistance between the two wire ends of the roll of AWG 12 copper wire shown.

(b) Using a K of 12.9, calculate the approximate length of the cable.

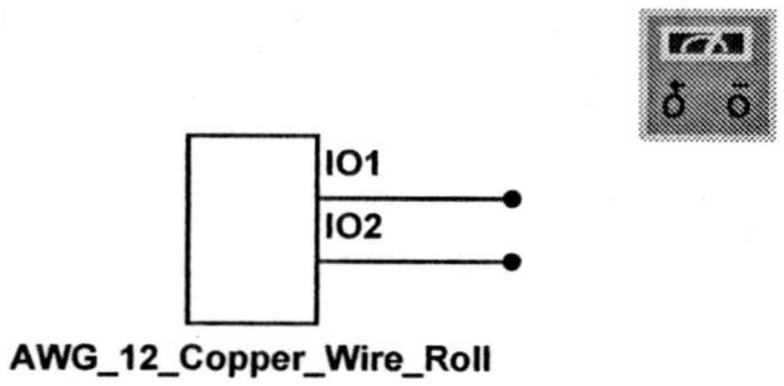

Answers

NAME ______________________________ DATE ________________

10-5 Open the Multisim circuit file CH10 _05 shown.

(a) Turn the simulation "on" and record the value of the load current.

(b) Calculate the approximate circuit voltage drop (use a K of 12.9).

(c) Calculate the value of the load voltage.

(d) Record the measured value of the load voltage.

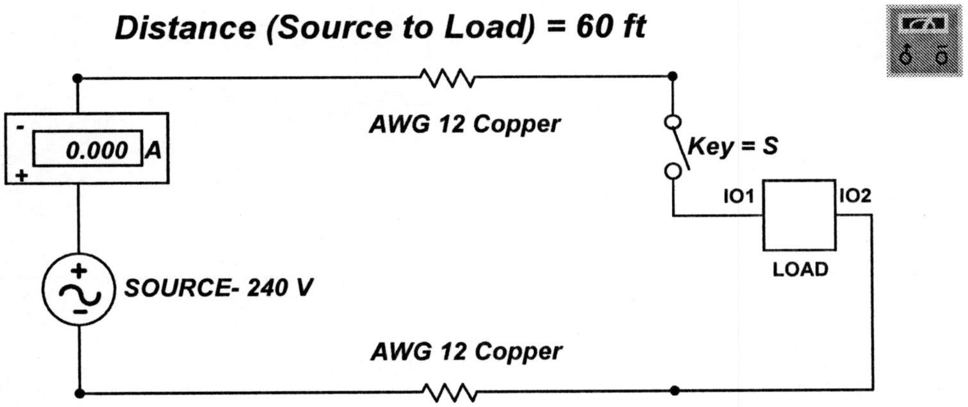

Answers

NAME ______________________ DATE ______________

10-6 Open the Multisim circuit file CH10 _06 shown.

(a) Turn the simulation "on" and record the operating voltage of the load.

(b) Repeat with an additional 50 ft of AWG 12 connected in series with each of the line wires of the original circuit.

(c) Repeat with an additional 50 ft of AWG 12 connected in parallel with each of the line wires of the original circuit.

(d) Which circuit (a, b, or c) has the lowest voltage drop? Why?

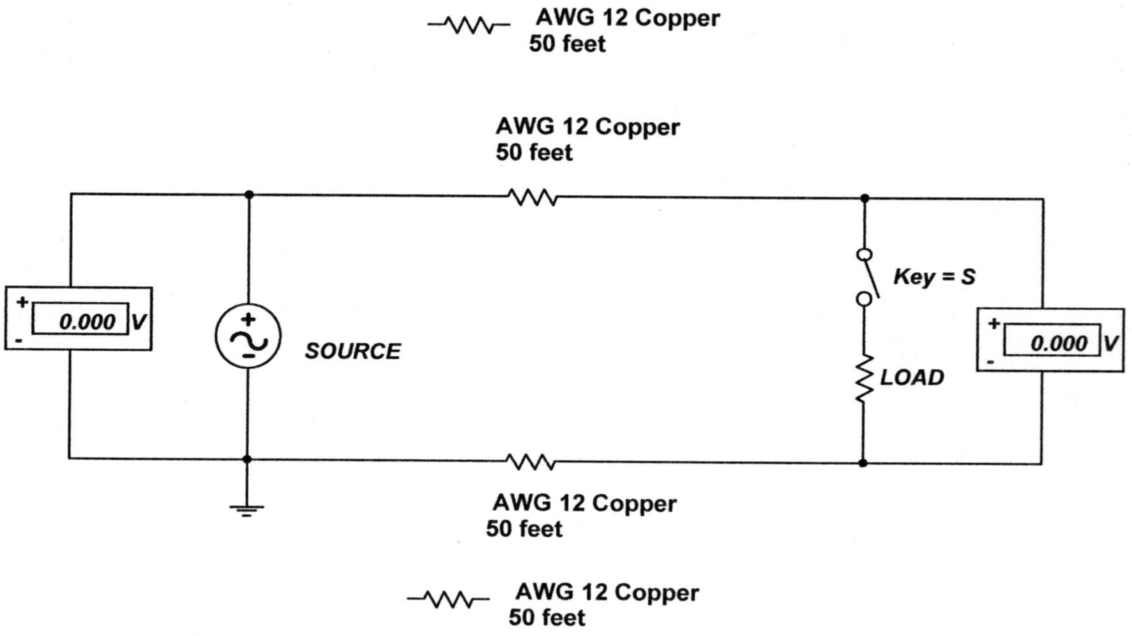

Answers

NAME ______________________________ DATE ________________

UNIT 11

RESISTORS

Multisim Simulation Assignments

11-1 Open the Multisim circuit file CH11 _01 shown.
Turn the simulation "on" and operate the circuit. Record the current for
(a) minimum (0 %) setting of the rheostat
(b) midpoint (50 %) setting of the rheostat
(c) maximum (100 %) setting of the rheostat

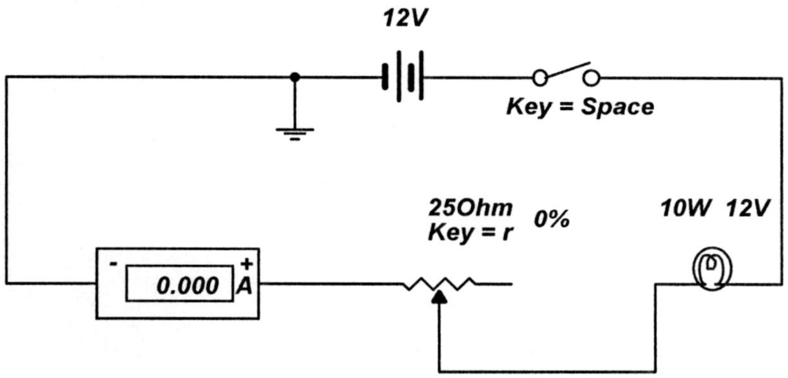

Press r to decrease % setting.
Press R to increase % setting.

<u>**Answers**</u>

NAME ______________________________ DATE ________________

11-2 Open the Multisim circuit file CH11 _02 shown. Turn the simulation "on" and operate the circuit. Record the voltage for
(a) minimum (0 %) setting of the potentiometer
(b) midpoint (50 %) setting of the potentiometer
(c) maximum (100 %) setting of the potentiometer

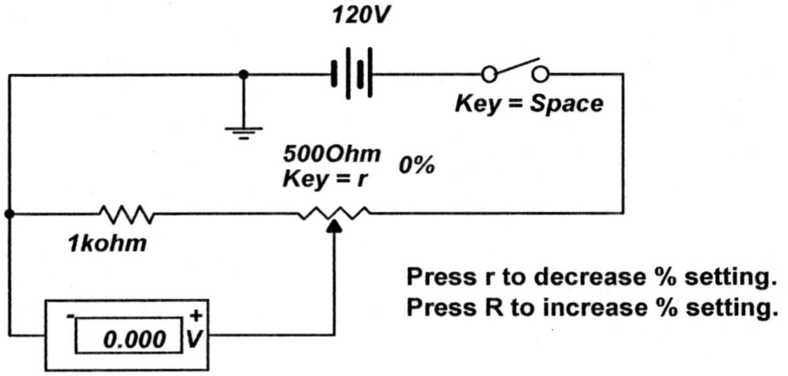

Answers

NAME ____ DATE ____________________

11-3 Open the Multisim circuit file CH11 _03 shown. Turn the simulation “on” and operate the circuit. Starting with the potentiometer at the zero resistance position (0 %) record the resistance reading for

(a) 1/4 turn (25 %) of the potentiometer
(b) 1/2 turn (50 %) of the potentiometer
(c) 3/4 turn (75 %) of the potentiometer
(d) full turn (100 %) of the potentiometer

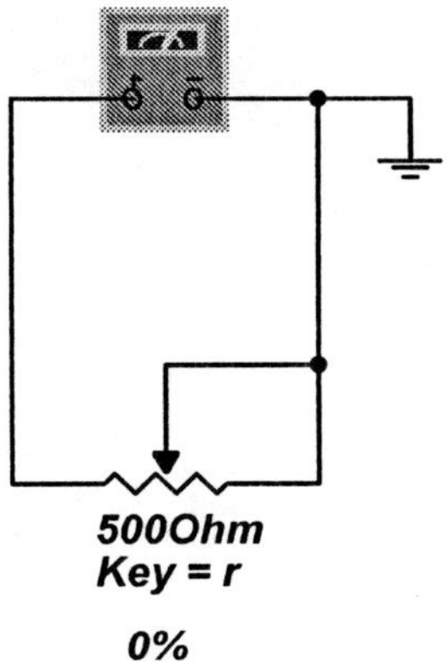

Press r to decrease % setting.
Press R to increase % setting.

Answers

NAME ____________________ DATE ____________

11-4 Open the Multisim circuit file CH11 _04 shown.

(a) Calculate the total resistance of the series connected resistor circuit shown.

(b) Turn the simulation "on" and measure the total resistance of the series connected resistor circuit shown.

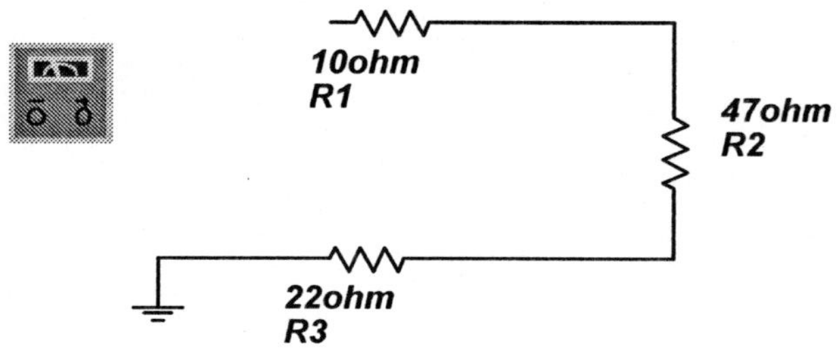

Answers

NAME ______________________ DATE ______________

11-5 Open the Multisim circuit file CH11 _05 shown.

(a) Calculate the voltages E1, E2, and E3 for the voltage-divider network shown.

(b) Turn the simulation "on" and measure the five different DC voltages available from this network.

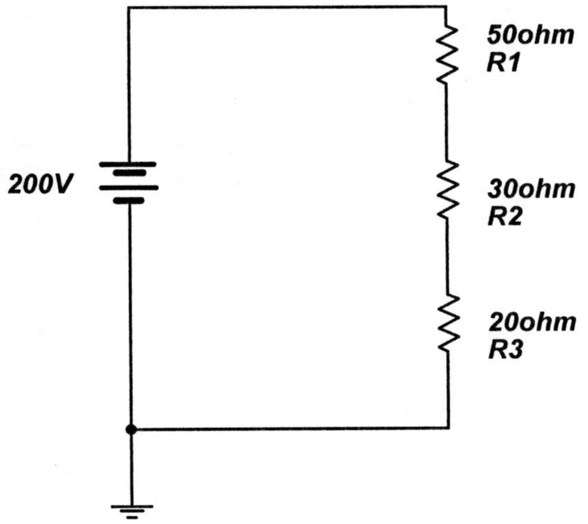

Answers

NAME ____________________ DATE ____________

11-6 Open the Multisim circuit file CH11 _06 shown.

(a) Calculate the total resistance for each circuit shown.

(b) Turn the simulation "on" and measure the total resistance of each circuit.

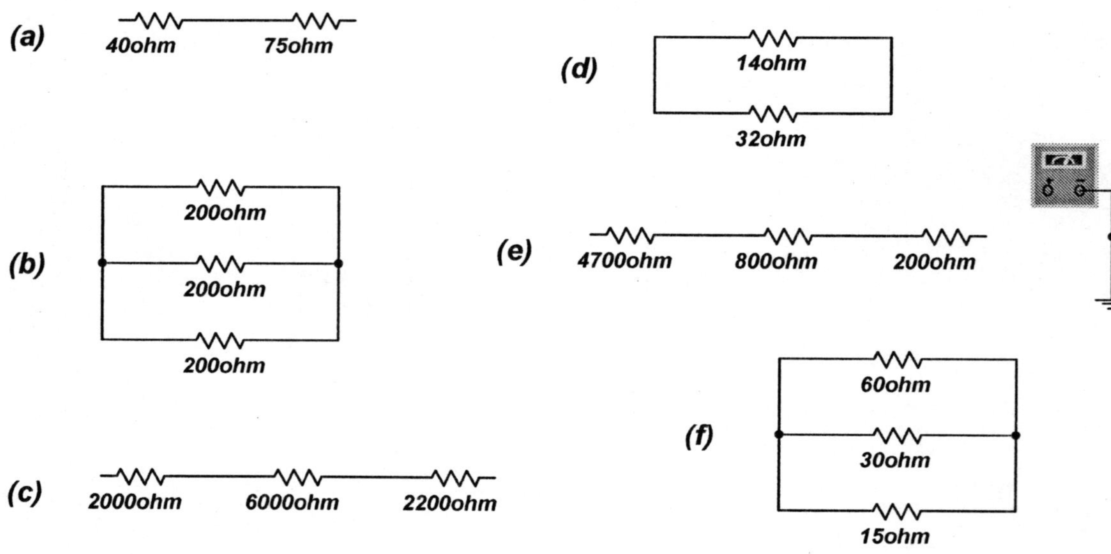

Answers

NAME ________________________________ DATE ____________________

11-7 Open the Multisim circuit file CH11 _07 shown.

(a) Calculate I_1 and I_2 for the current-divider circuit shown.

(b) Turn the simulation "on" and use the multimeter to measure the value of I_1 and I_2.

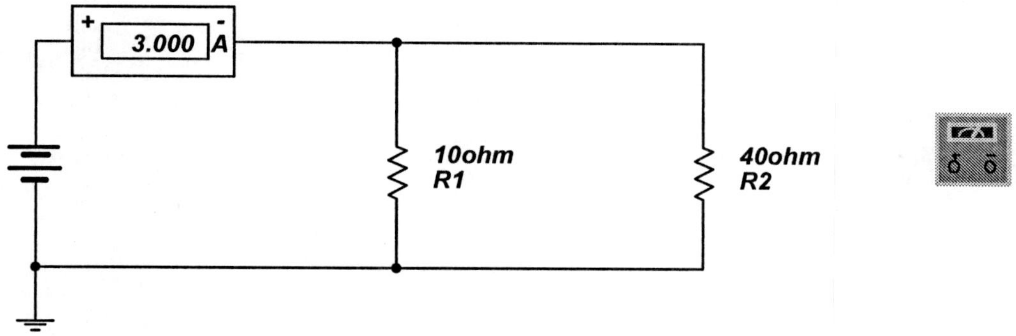

Answers

NAME ______________________________ DATE ________________

11-8 Open the Multisim circuit file CH11 _08 shown.

(a) Calculate the total resistance of the series-parallel connected resistor circuit shown.

(b) Turn the simulation "on" and measure the total resistance of the series-parallel connected resistor circuit shown.

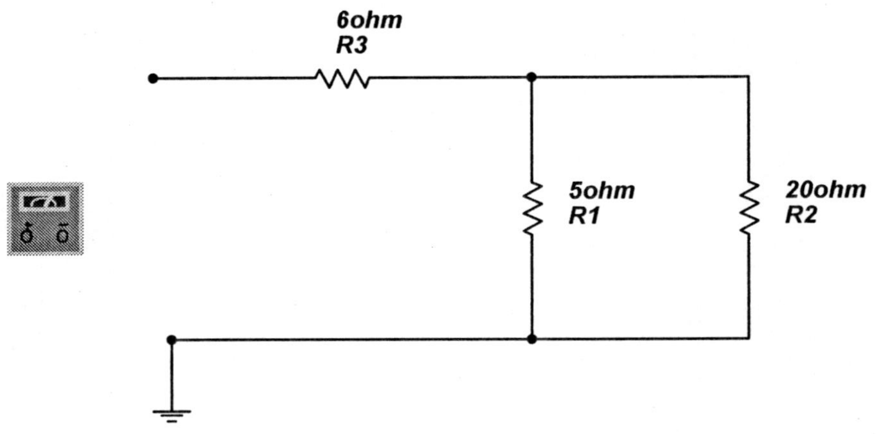

Answers

NAME ____________________ DATE ____________

11-9 Open the Multisim circuit file CH11 _09 shown.

(a) You have been given three 100 Ω resistors to connect together in three different circuit configurations. State the name of each of the possible circuit configurations and the total resistance expected for each.

(b) Connect each circuit in turn and record the total resistance as measured with the multimeter.

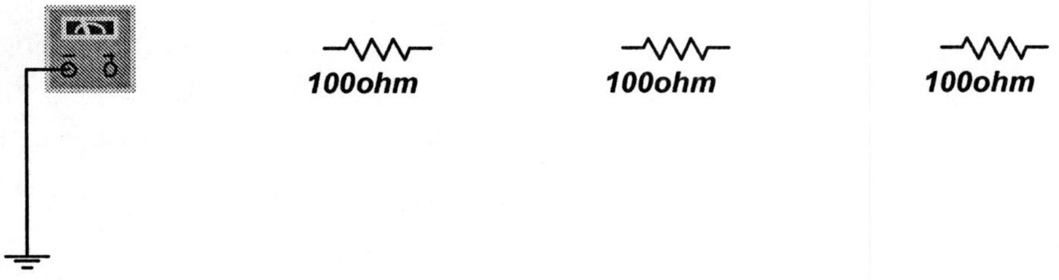

Answers

NAME ______________________________ DATE ________________

11-10 Open the Multisim circuit file CH11 _10 shown.

(a) Calculate the total resistance of the resistor network shown. Turn the simulation "on" and verify your answer by measuring the total resistance.

(b) Assume resistor R_1 burnt open (infinite resistance). Calculate the new value of the total resistance. Turn the simulation "on" and verify your answer by replacing R_1 with the open resistor and measuring the total resistance.

(c) Assume resistor R_3 of the original circuit is short-circuited (zero resistance). Calculate the new value of the total resistance. Turn the simulation "on" and verify your answer by replacing R_3 with the shorted resistor and measuring the total resistance.

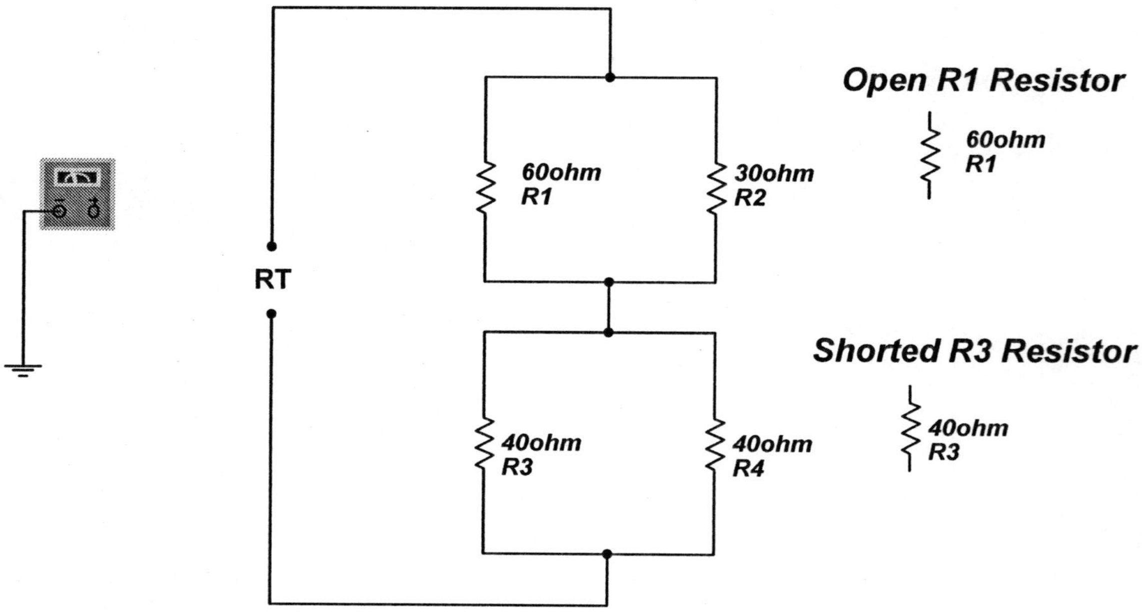

Answers

NAME ______________________________ DATE ________________

UNIT 12

ELECTRICITY AND MAGNETISM

Multisim Simulation Assignments

12-1 Open the Multisim circuit file CH12 _01 shown.
Turn the simulation "on" and use the multimeter to measure:
(a) the AC input voltage applied across the primary
(b) the AC output voltage obtained across the secondary coil.

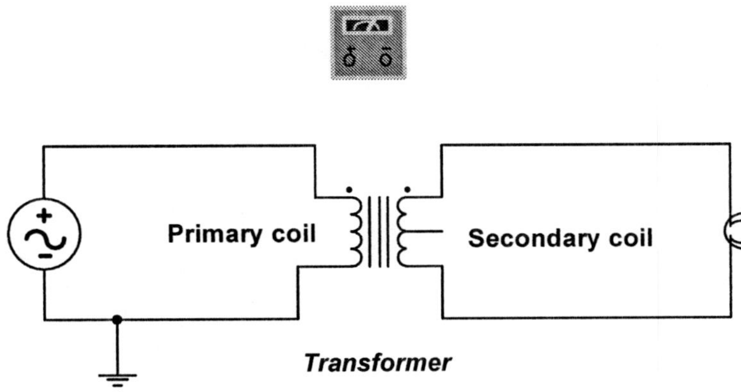

Answers

NAME ______________________________ DATE ______________

12-2 Open the Multisim circuit file CH12 _02 shown.
Turn the simulation "on".
(a) Close the switch to establish current flow through the relay coil and make note of what happens.
(b) Complete the wiring of the contact to control the alarm device.
Use the multimeter to record the AC operating current of the alarm.

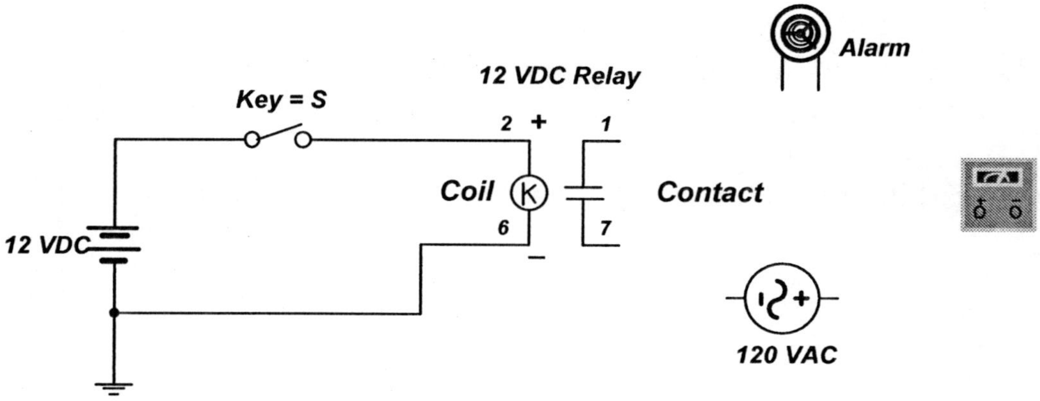

Answers

NAME ______________________________ DATE ________________

UNIT 13

RELAYS

Multisim Simulation Assignments

13-1. Open the Multisim circuit file CH13 _01 shown. Complete a wiring diagram of the circuit that uses the relay to control the 120 VAC lamp circuit with the 12 VDC control circuit.
Turn the simulation "on" and verify your circuit operation by running the simulation.

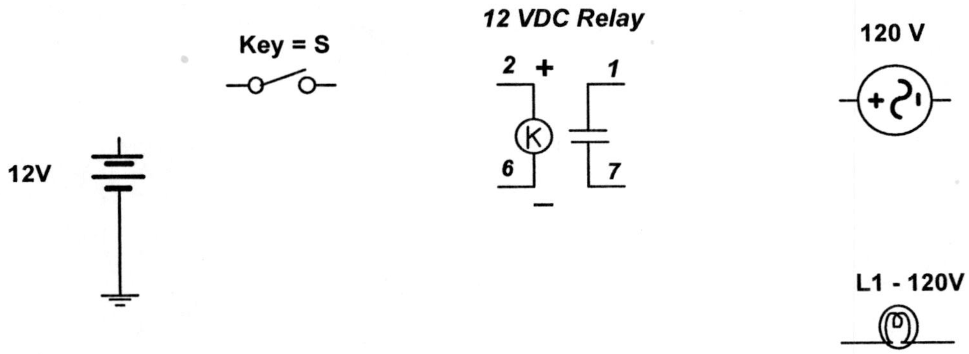

Answers

NAME ______________________________ DATE ____________________

13-2. Open the Multisim circuit file CH13 _02 shown. Turn the simulation "on" and use the multimeter to measure:
(a) relay coil control current.
(b) starter motor current.

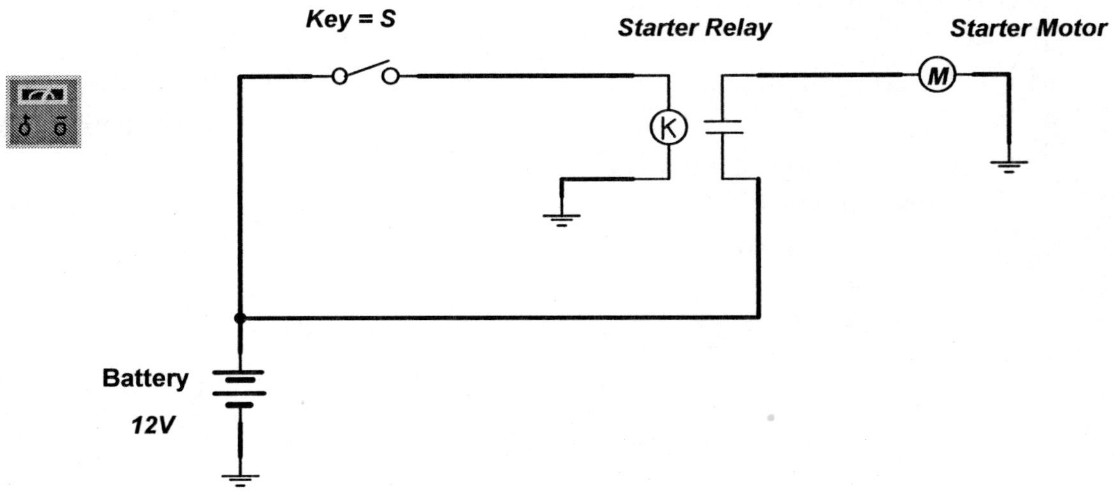

Answers

NAME ______________________________ DATE ______________

13-3. Open the Multisim circuit file CH13 _03 shown. Turn the simulation "on". Operate the circuit and answer the following:
(a) What type of timer circuit is simulated?
(b) What is the approximate length of the time delay period?

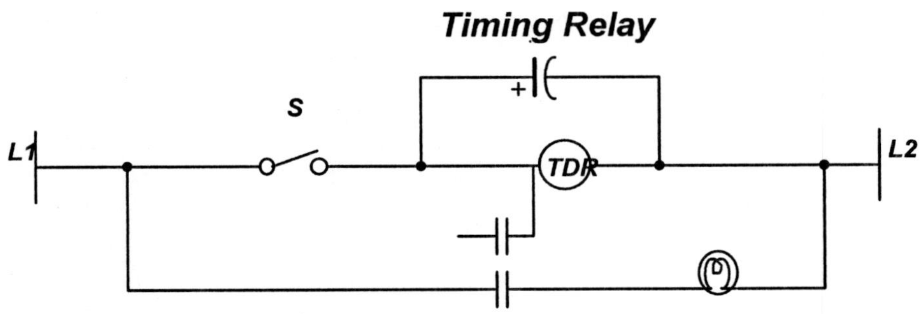

Answers

NAME ____________________ DATE ____________

13-4. Open the Multisim circuit file CH13 _04 shown. Turn the simulation "on" and check the contacts of the relay using the ohmmeter to determine which if any of the contacts are bad. Give a reason for your answer.

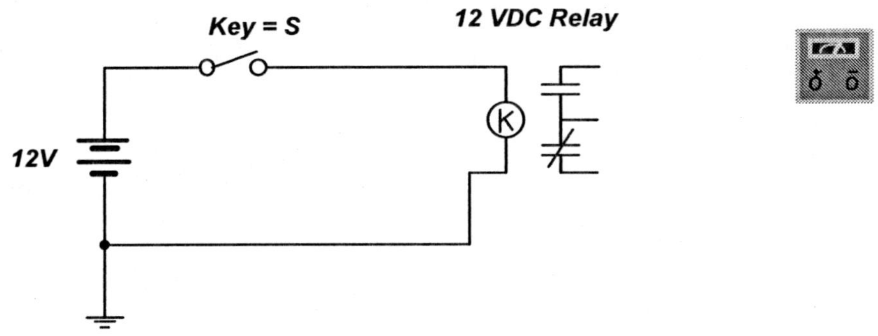

Answer

NAME ______________________ DATE ______________

13-5. Open the Multisim circuit file CH13 _05 shown. Assume a third light is to be connected to the circuit to come "on" whenever the relay coil is energized. At what point in the circuit would this light have to be connected? Redraw the circuit schematic with this light properly connected.
Add the third light to the circuit, turn the simulation “on” and verify that is properly connected.

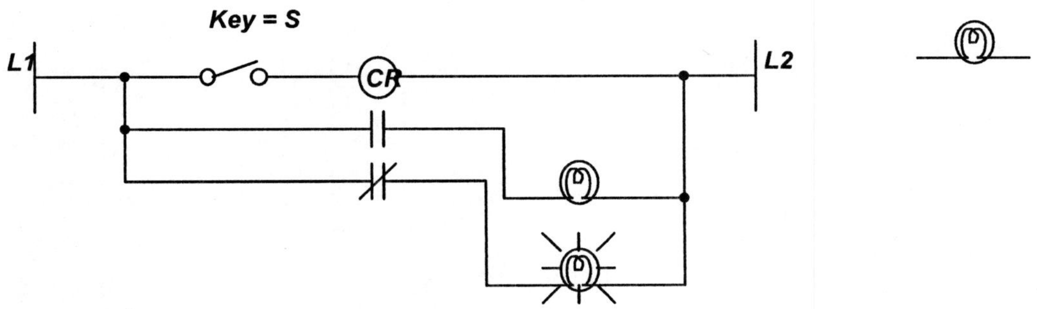

Answers

NAME ___________________________________ DATE _________________

13-6. Open the Multisim circuit file CH13 _06 shown. Complete a wiring diagram of the circuit with the relay contacts connected to operate the lights as follows:
- with switch "S" open L1 is "on" and L2 is "off"
- with switch "S" closed L1 is "off" and L2 is "on"

Wire the circuit, turn the simulation "on" and verify that is properly connected.

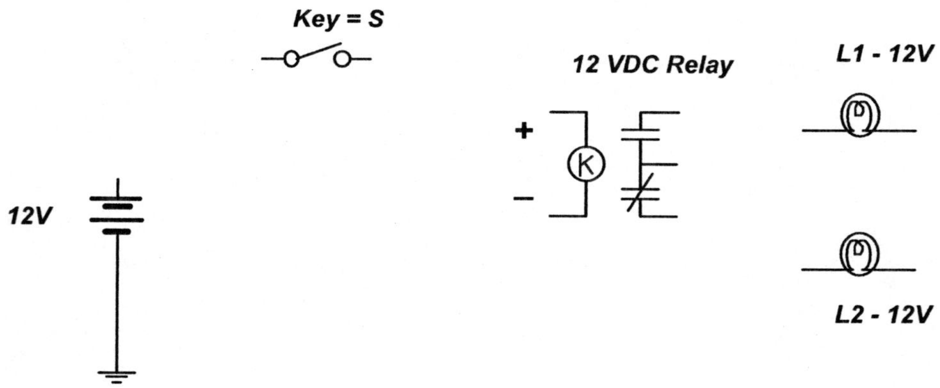

Answer

NAME ____________________ DATE ____________

UNIT 14

CIRCUIT PROTECTION DEVICES

Multisim Simulation Assignments

14-1. Open the Multisim circuit file CH14 _01 shown.
 (a) Turn the simulation "on" and record the current drawn by each lamp load.
 (b) Closing what two switches will result in an overload condition? What is the value of this overload current?
 (c) What three switches can be closed without creating an overload condition?
 What is the value of this total current flow?

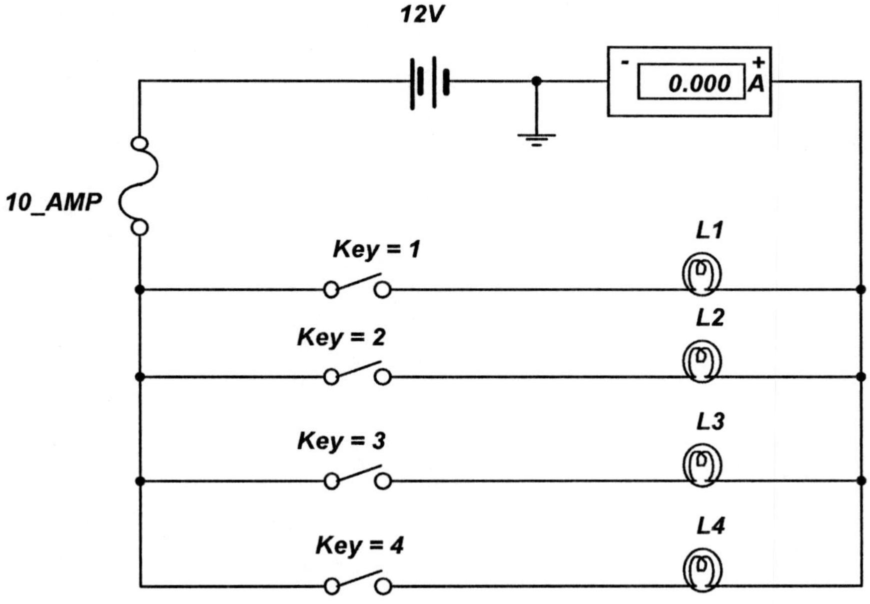

Answers

NAME ______________________________ DATE ________________

14-2. Open the Multisim circuit file CH14 _02 shown and turn the simulation "on". Closing what switch(s) will create a short circuit?

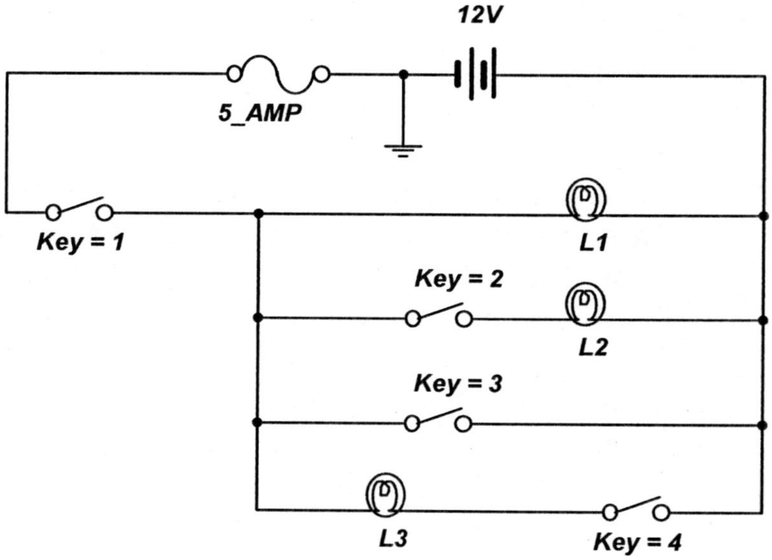

Answer

NAME ______________________________ DATE ________________

14-3. Open the Multisim circuit file CH14 _03 shown and turn the simulation "on".
(a) Record the voltage on the LINE side of the fuses.
(b) Record the voltage on the LOAD side of the fuses.
(c) Record the voltage from L1 to T2.
(d) Record the voltage from L2 to T1.
(e) Base on your readings, which fuse, is burnt open? (Replace blown fuse to confirm your answer)

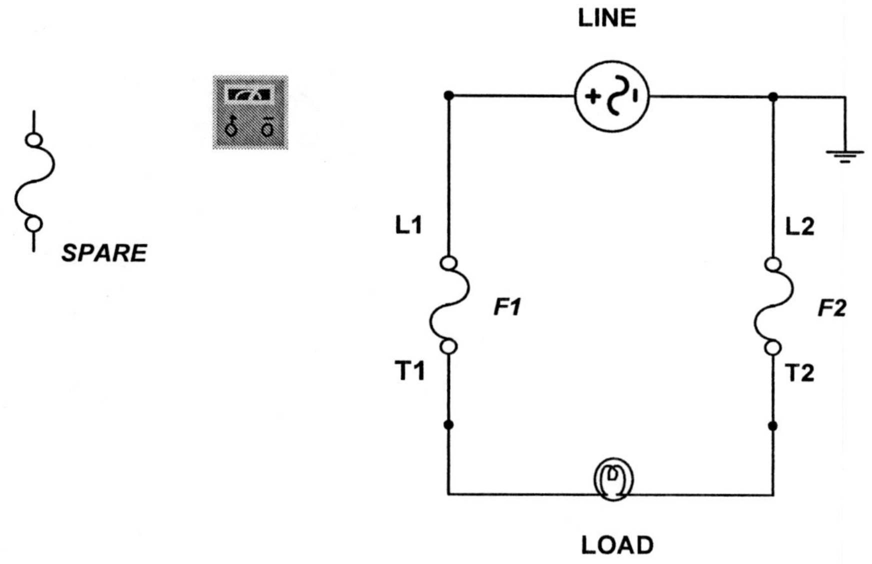

Answers

NAME ______________________________ DATE ______________

14-4. Open the Multisim circuit file CH14 _04 shown and turn the simulation "on". Use the multimeter to record the resistance of the GOOD and the burnt OPEN fuse.

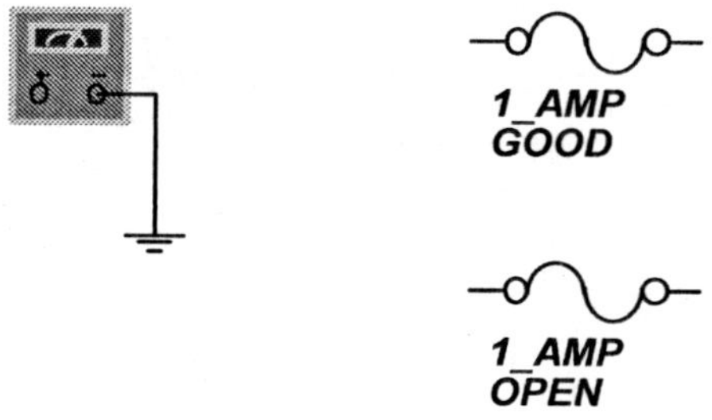

Answer

NAME ______________________________________ DATE ____________________

14-5. Open the Multisim circuit file CH14 _05 shown and turn the simulation "on". Use the multimeter to determine which fuses are good and which are bunt open.

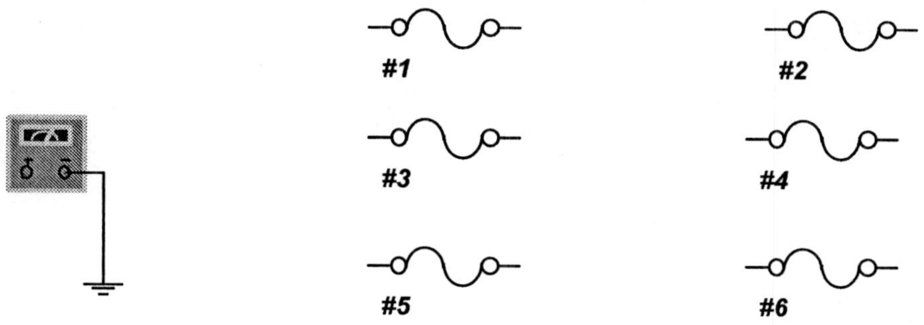

Answers

NAME ______________________ DATE ______________

14-6. Open the Multisim circuit file CH14 _06 shown and turn the simulation "on".

(a) Measure the voltage drop across the fuse for a normally operating circuit.

(b) Repeat with the open fuse connected in the circuit.

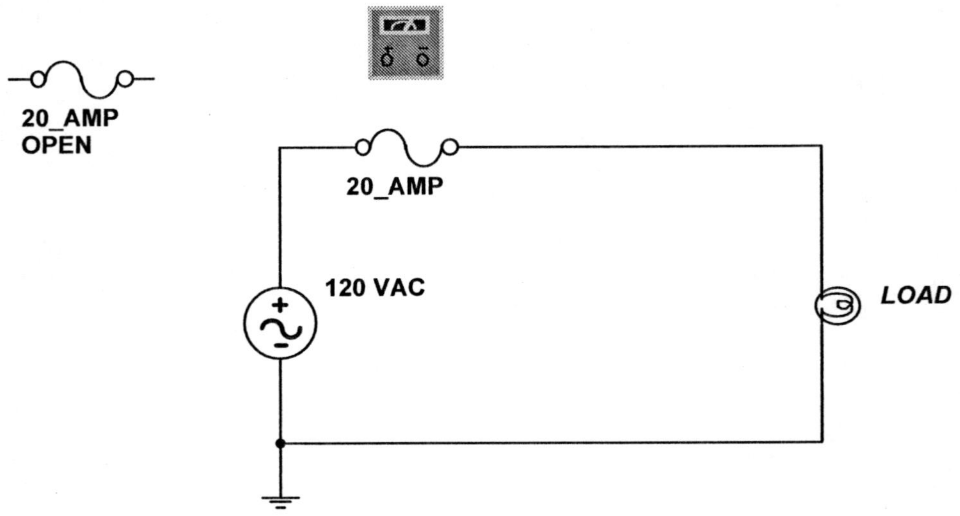

Answers

NAME ______________________________ DATE ________________

UNIT 15

ELECTRIC POWER AND ENERGY

Multisim Simulation Assignments

15-1. Open the Multisim circuit file CH15 _01 shown.
  (a) Turn the simulation "on" and record the current drawn by the range element.
  (b) Calculate the power rating of the range element.
  (c) Connect the wattmeter and measure the amount of power being used by the range element.

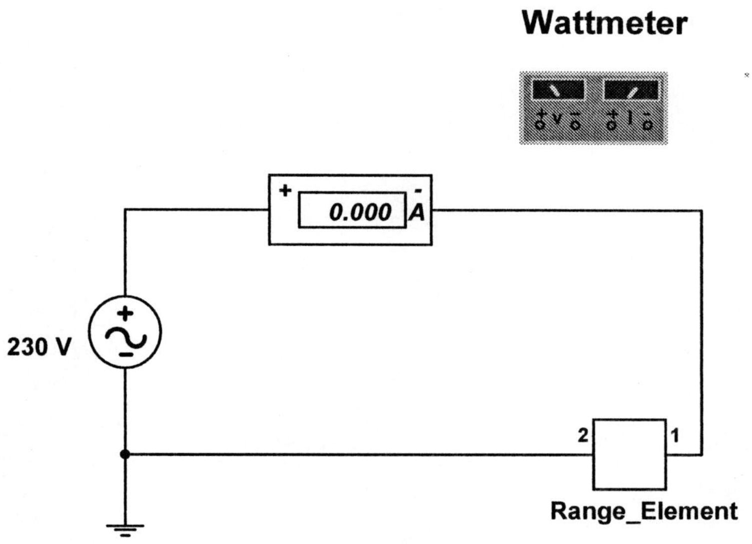

Answers

NAME ______________________ DATE ______________

15-2. Open the Multisim circuit file CH15 _02 shown.

(a) Turn the simulation "on" and record the amount of power being converted by the toaster.

(b) Calculate the amount of current being drawn by the toaster.

(c) Connect the multimeter and measure the amount of current being drawn by the toaster.

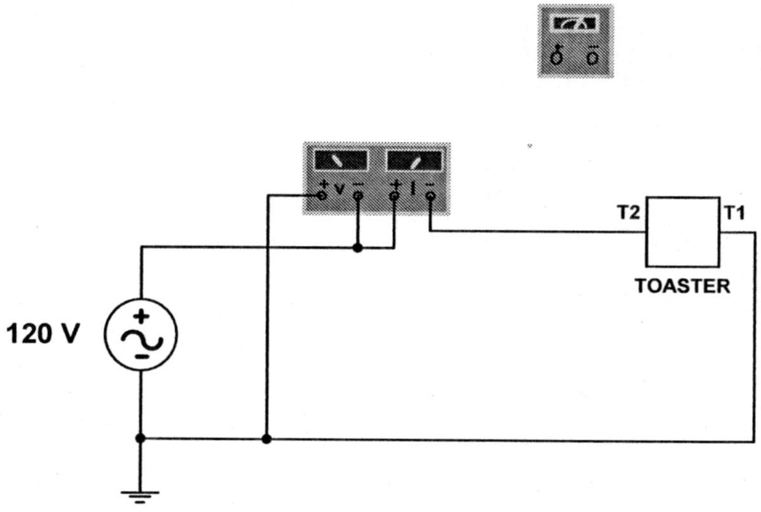

Answers

NAME ______________________________ DATE ______________

15-3. Open the Multisim circuit file CH15 _03 shown.
  (a) Turn the simulation "on" and record the amount of current being drawn by the resistor.
  (b) Calculate the amount of power being dissipated in the form of heat from the resistor.
  (c) Connect the wattmeter and measure the amount of power being dissipated by the resistor.

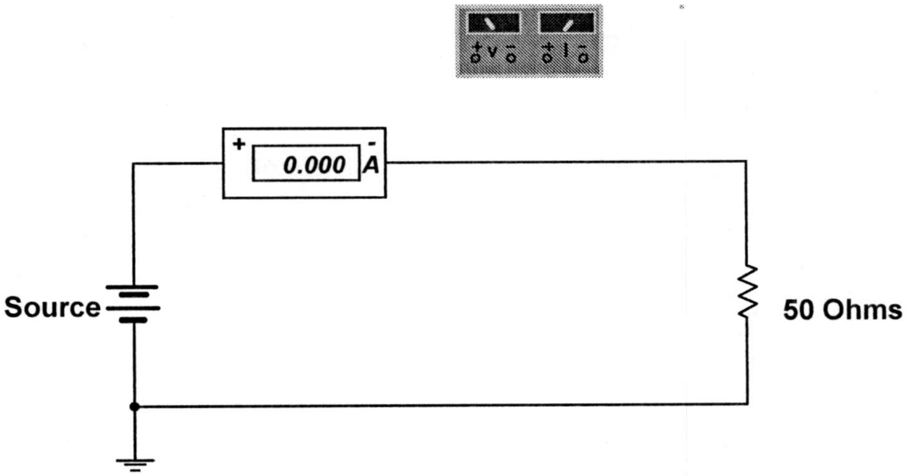

Answers

NAME ____________________ DATE ____________

15-4. Open the Multisim circuit file CH15 _04 shown.

(a) Calculate the value of the operating current for each of the 120V bulbs shown.

(b) Turn the simulation "on" and use the multimeter to measure the operating current of each bulb.

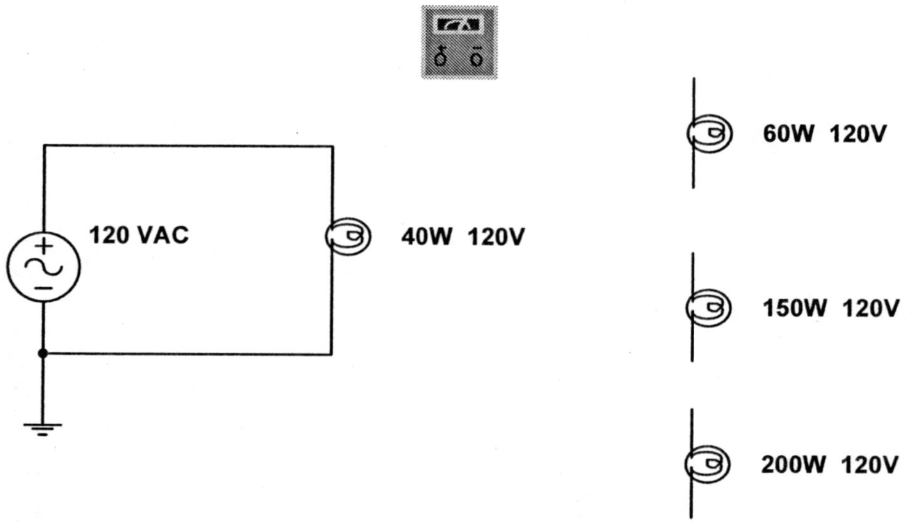

Answers

15-5. Open the Multisim circuit file CH15 _05 shown.
(a) Turn the simulation "on" and record the voltmeter and ammeter readings.
(b) Calculate the total power of the circuit.
(c) Connect the wattmeter and measure the total power of the circuit.

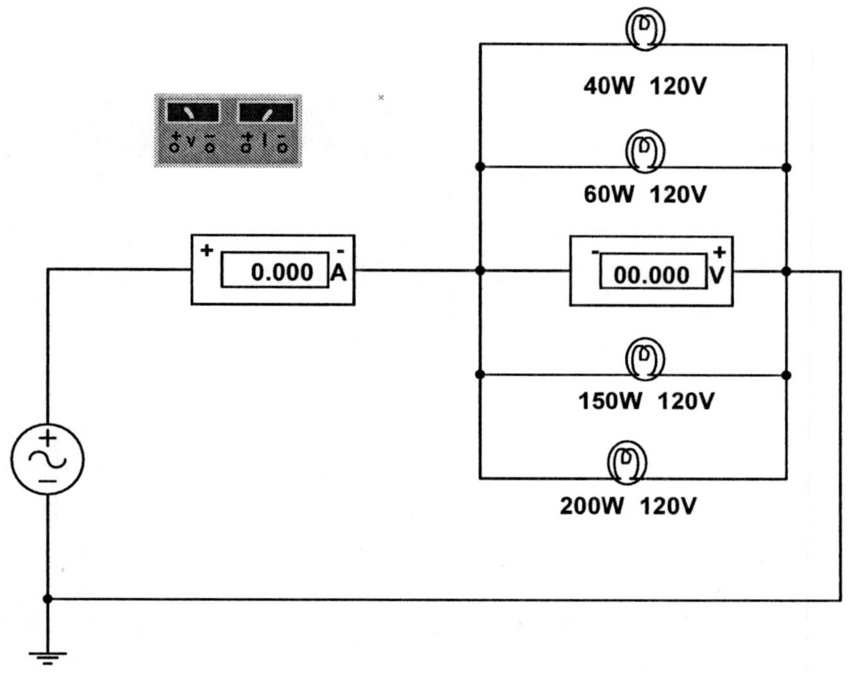

Answers

NAME ________________________________ DATE ________________

UNIT 16

OVERVIEW OF ELECTRIC CIRCUITS AND SYSTEMS

Multisim Simulation Assignments

16-1. Open the Multisim circuit file CH16 _01 shown.
The front chime does not operate when the front pushbutton is pressed. The rear chime operates fine.
(a) What components would normally be eliminated as a possible cause of the problem? Why?
(b) Turn the simulation "on" and use the multimeter to determine which component is at fault. Replace the defective component with one of the spare parts.

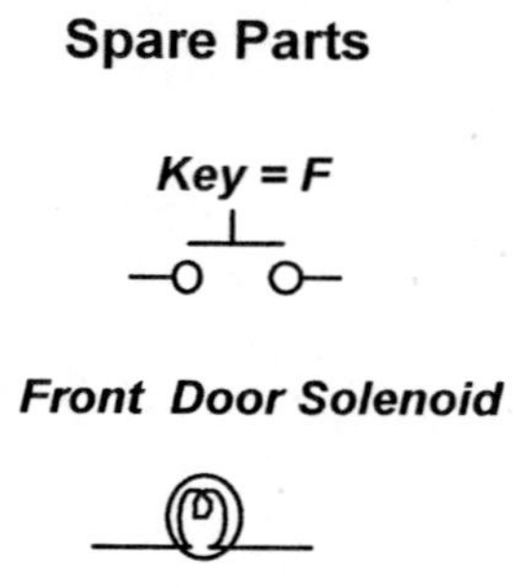

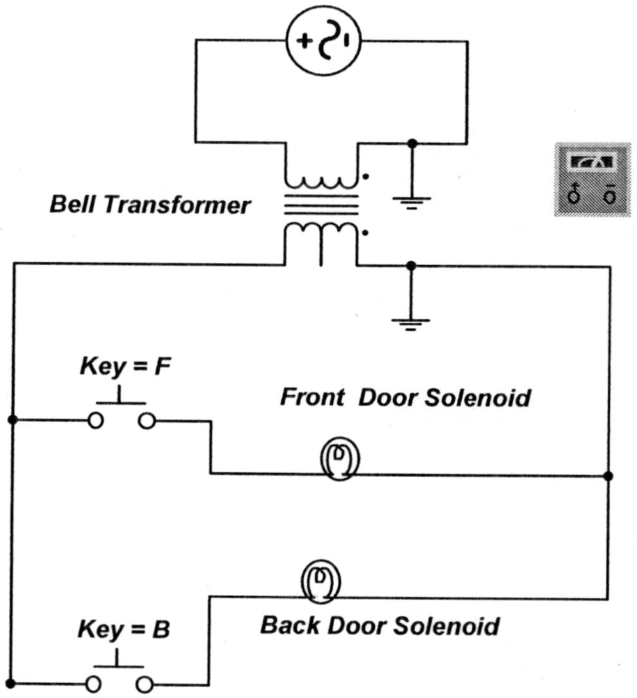

Answers

NAME ____________________ DATE ____________

16-2. Open the Multisim circuit file CH16 _02 shown.
Properly connect each of the three loads (light, heater, and dryer) to the 3-wire distribution system. Turn the simulation "on" and measure the amount of current drawn by the:
(a) light
(b) heater
(c) dryer motor
(d) dryer heating element

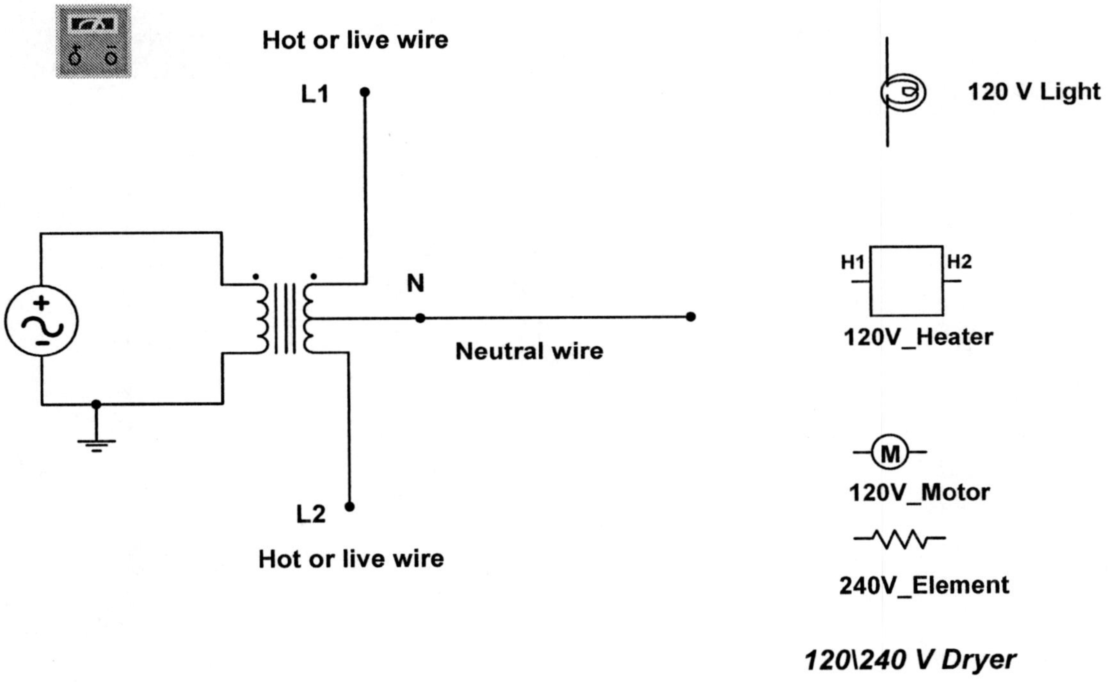

Answers

NAME ________________________________ DATE ________________

16-3. Open the Multisim circuit file CH16 _03 shown. Turn the simulation "on" and for the balanced 3-wire distribution system shown (Load L1 same as Load L2) record the value of the:

(a) (i) transformer primary voltage
(ii) voltage from any one of the hot wires to neutral
(iii) voltage across the two hot wires
(iv) current flow through L1
(v) current flow through L2
(vi) current flow through the neutral

(b) Repeat with L2 replaced by L3 to simulate an unbalanced load condition.

(c) What is the relationship of the current flow in the neutral compared to that in the two hot wires?

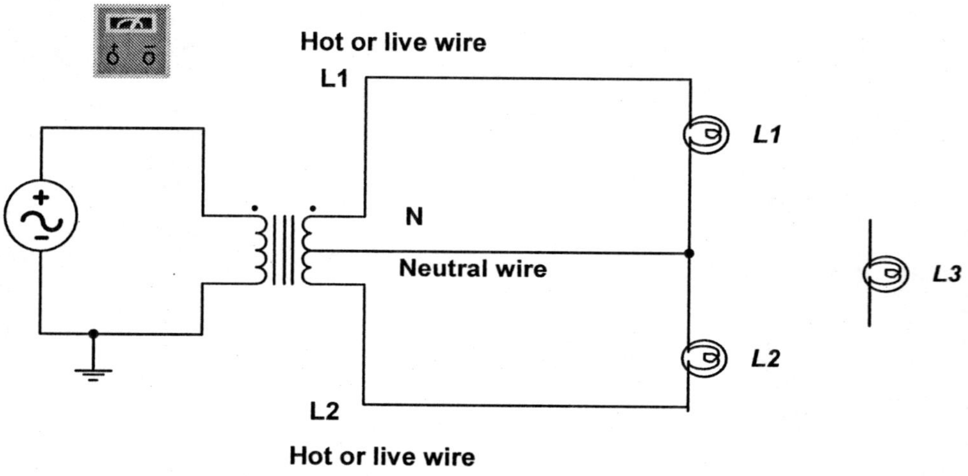

Answers

NAME ___________________________ DATE ________________

16-4. Open the Multisim circuit file CH16 _04 shown. Properly connect each the six 120V branch circuits (current for each is as specified) to the service panel. Distribute the loads so that MINIMUM current flows through the neutral wire. Record:
(a) the amount of current flow through each of the hot wires
(b) the amount of current flow through the neutral wire

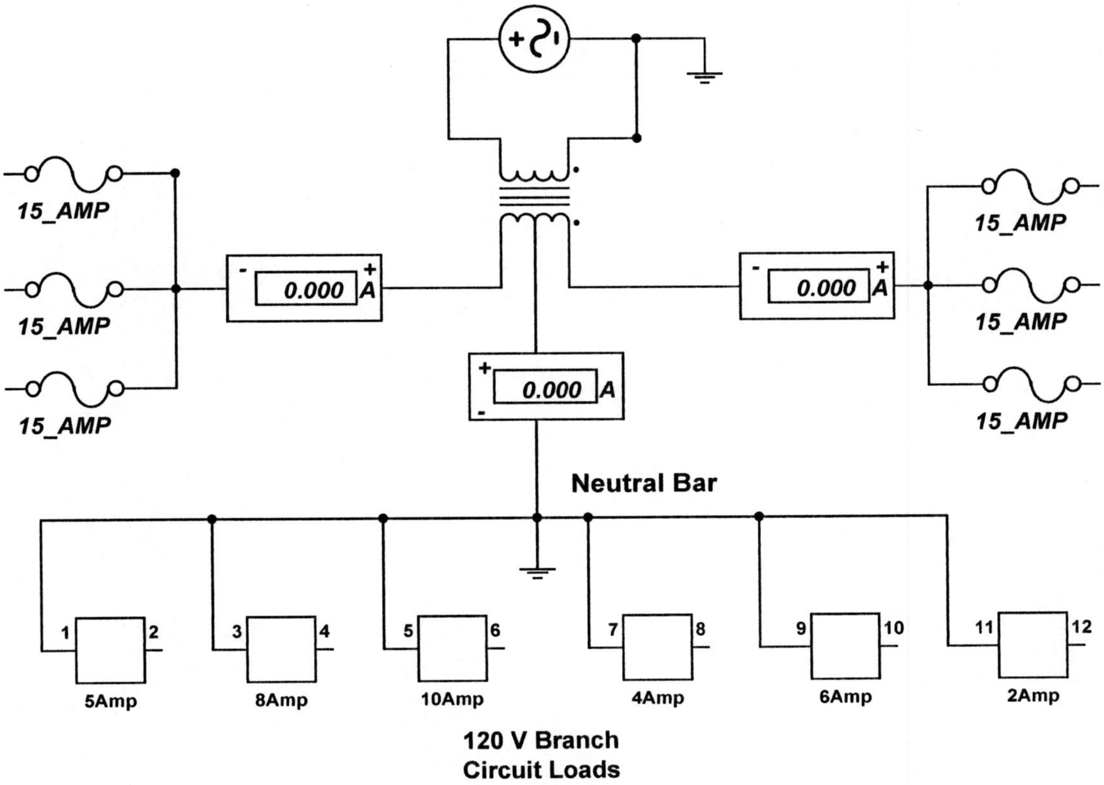

Answers

NAME ____________________ DATE ____________

16-5. Open the Multisim circuit file CH16 _05 shown. Wire the circuit, using the components in the positions shown, to simulate the operation of a single-pole, single-throw, switch used to control two lights from one position.

Answer – Complete the wiring diagram

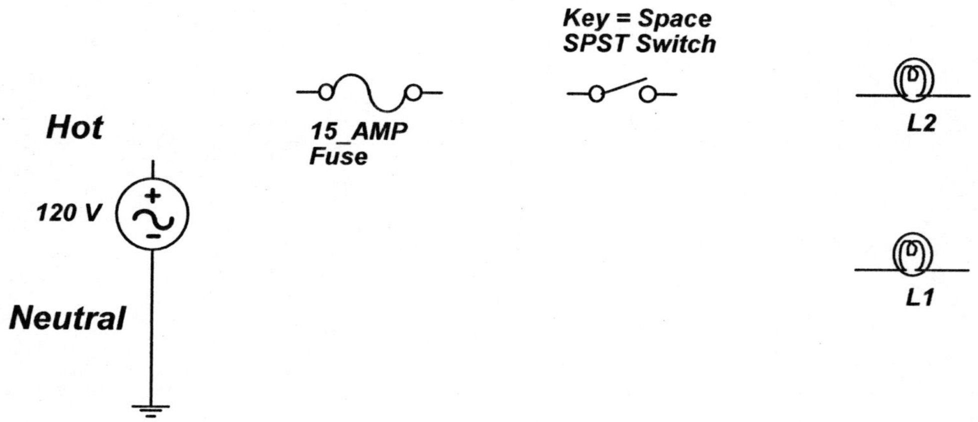

NAME ________________________________ DATE ________________

16-6. Open the Multisim circuit file CH16 _06 shown. Wire the circuit, using the components in the positions shown, to simulate the operation of a two 3-way switches used to control two lights from two positions.

Answer – Complete the wiring diagram

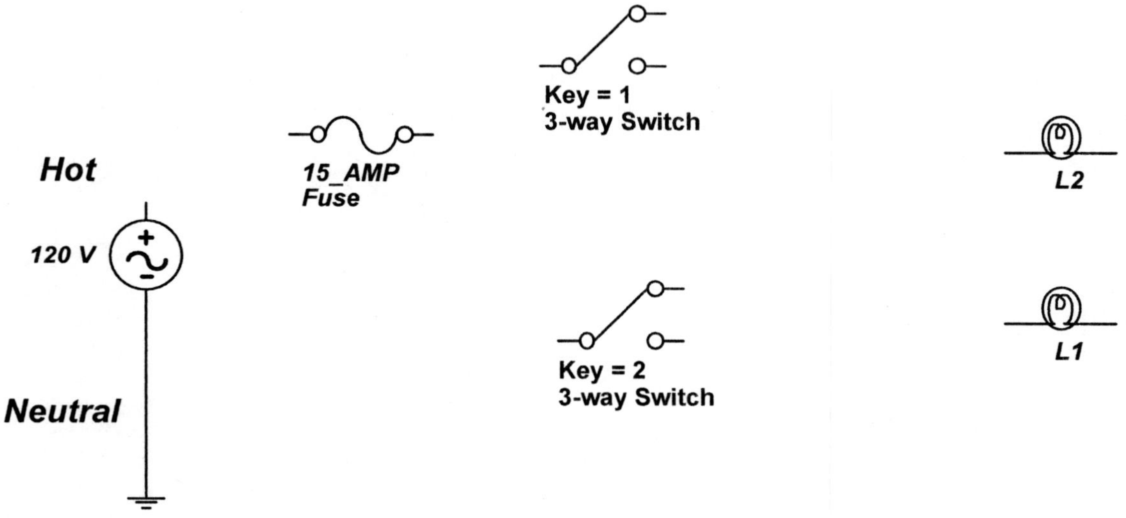

NAME ______________________________________ DATE ____________________

16-7. Open the Multisim circuit file CH16 _07 shown. Turn the simulation "on" and operate the circuit. Note that the fuse blows as soon as the switch is closed. Troubleshoot the circuit with the aid of the multimeter to determine the problem.

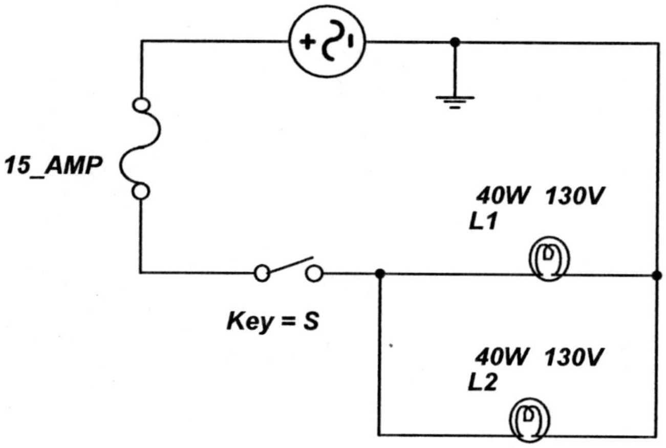

Answer

NAME ______________________________ DATE ________________

16-8. Open the Multisim circuit file CH16 _08 shown. Turn the simulation "on" and operate the circuit. What problem, if any, exists?

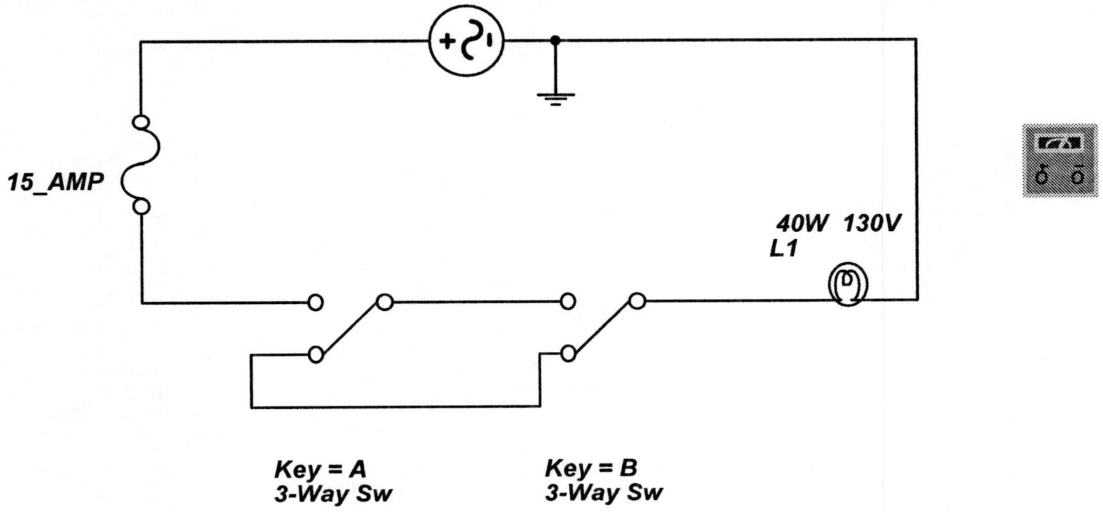

Answer

NAME ________________________________ DATE ________________

16-9. Open the Multisim circuit file CH16 _09 shown. Turn the simulation "on", operate the circuit and answer each of the following:
(a) What is the value of the normal operating current of the appliance?
(b) What is the value of the normal case-to-ground voltage?
(c) What is the value of the case-to-ground voltage with the case ground switch open and the ground fault switch closed?
(d) What happens when both the ground fault and appliance ground switches are closed?

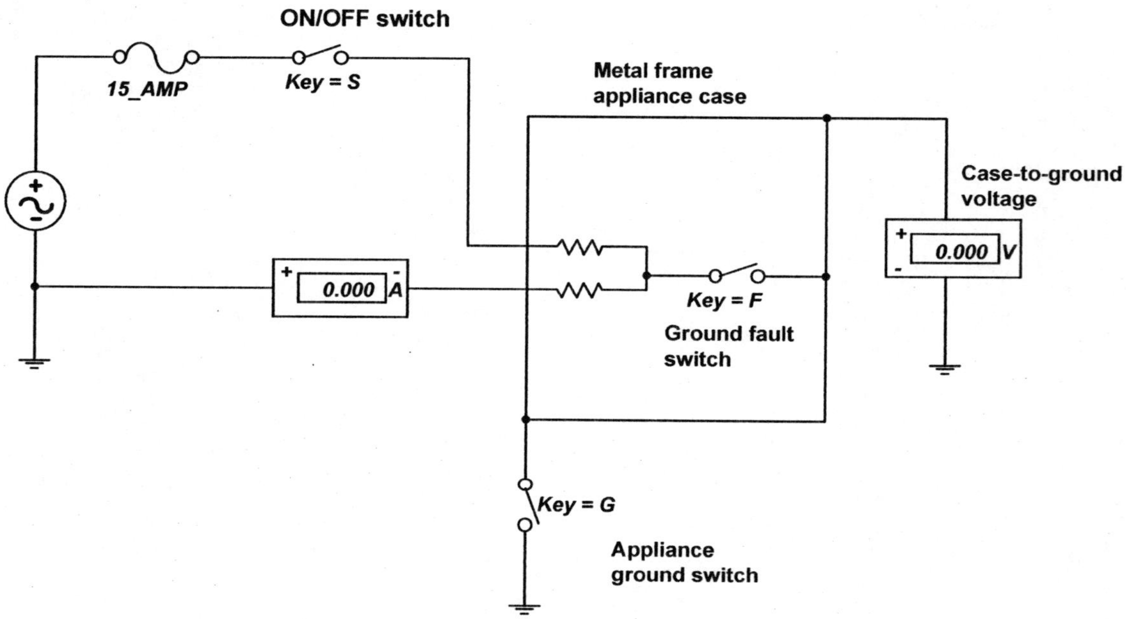

Answers

NAME ______________________________ DATE __________________

UNIT 17

SOLVING THE DC SERIES CIRCUIT

Multisim Simulation Assignments

17-1. Open the Multisim circuit file CH17 _01 shown. Use the multimeter for all voltage, current, and resistance measurements. Use the wattmeter for all power measurements.

(a) Measure the source voltage and the voltage across each resistor.

(b) Based on your readings, state the voltage characteristics of a series circuit.

(c) Measure the source current and the current flow through each resistor.

(d) Based on your current readings state the current characteristics of a series circuit.

(e) Measure the total circuit resistance and the resistance of each resistor.

(f) Based on your resistance readings state the resistance characteristics of a series circuit.

(g) Measure the power dissipation of each resistor and that of the total circuit.

(h) Based on your wattmeter readings state the power characteristics of a series circuit.

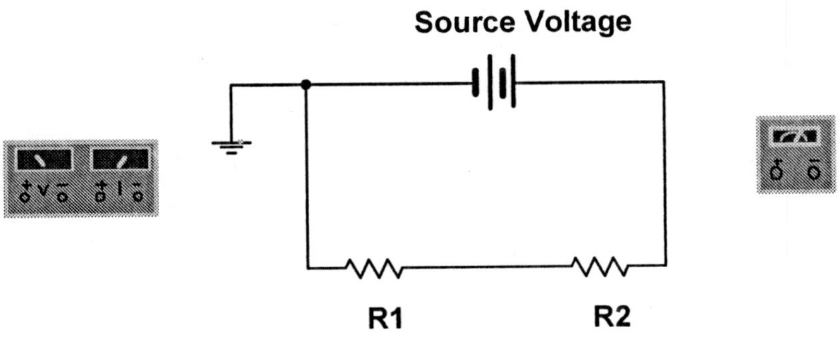

Answers

NAME ______________________________ DATE ________________

17-2. Open the Multisim circuit file CH17 _02 shown. Use the multimeter for all voltage, current, and resistance measurements.

(a) Calculate the value of the total resistance and measure the total resistance to verify your answer.

(b) Calculate the value of the current flow and measure the current flow to verify your answer.

(c) Calculate the value of the voltage across R_1 and R_2 and measure the their voltage drops to verify your answer.

(d) Why is the voltage drop across R_1 equal in value to that across R_2?

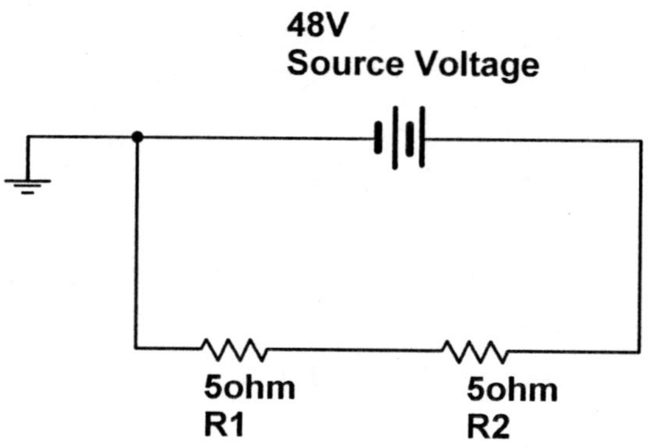

Answers

17-3. Open the Multisim circuit file CH17 _03 shown. Use the multimeter for all voltage, current, and resistance measurements.

(a) Calculate the value of the total resistance and measure the total resistance to verify your answer.

(b) Calculate the value of the current flow and measure the current flow to verify your answer.

(c) Calculate the value of the voltage across R_1 and R_2 and measure the their voltage drops to verify your answer.

(d) Why is the voltage drop across R_2 greater than that across R_1?

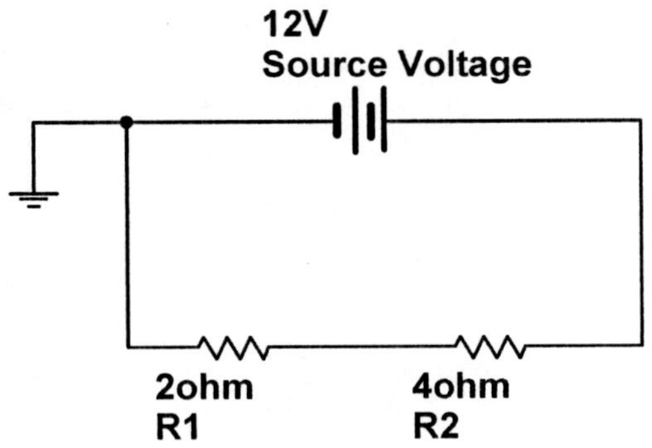

Answers

NAME ______________________________ DATE ________________

17-4. (a) Open the Multisim circuit file CH17 _04 shown. Find all the unknown values of E, I, R, and P, for the series circuit. Record the all given and calculated values in the table provided.

(b) Use the multimeter and wattmeter to measure all values of E, I, R, P, and record your readings in the table provided.

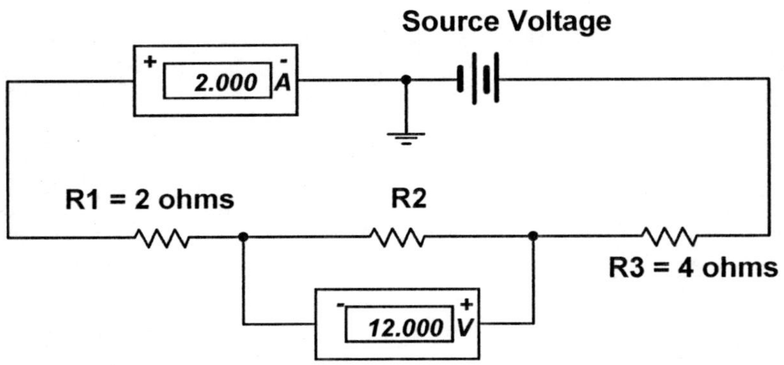

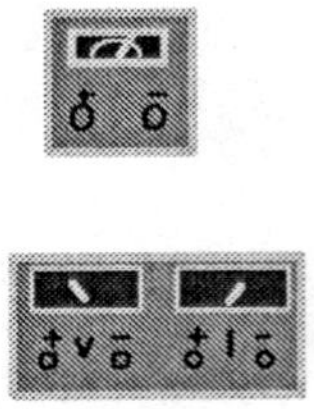

Answers

(a)

	Voltage	Current	Resistance	Power
R_1				
R_2				
R_3				
Total				

(b)

	Voltage	Current	Resistance	Power
R_1				
R_2				
R_3				
Total				

NAME ______________________________ DATE ________________

17-5. (a) Open the Multisim circuit file CH17 _05 shown. Find all the unknown values of E, I, and R, for the series circuit. Record the all given and calculated values in the table provided.

(b) Use the multimeter to measure all values of E, I, R, and record your readings in the table provided.

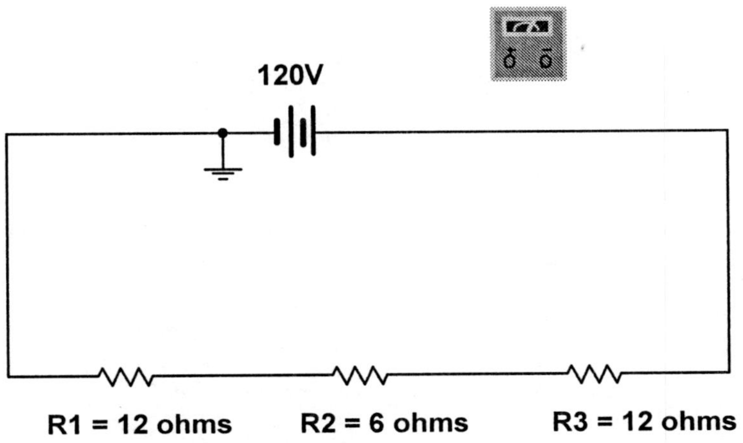

Answers

(a)

	Voltage	**Current**	**Resistance**
R_1			
R_2			
R_3			
Total			

(b)

	Voltage	**Current**	**Resistance**
R_1			
R_2			
R_3			
Total			

NAME ____________________________________ DATE __________________

17-6. Open the Multisim circuit file CH17 _06 shown. Using the multimeter record the polarity (positive or negative) and voltage from:
(a) point A to common ground
(b) point B to common ground
(c) point C to common ground

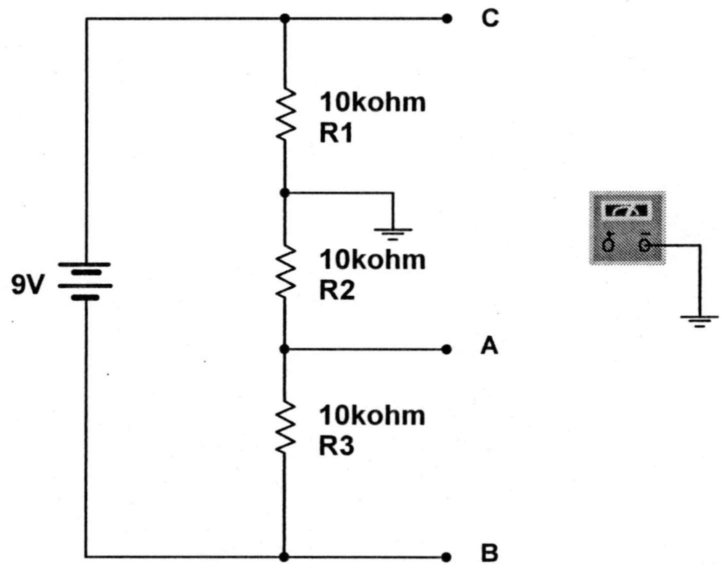

Answers

NAME ______________________________ DATE ________________

17-7. Open the Multisim circuit file CH17 _07 shown.
(a) Calculate the value of the current through and the voltage across the load with the batteries connected series-aiding. Turn the simulation "on" and verify your calculations.
(b) Calculate the value of the current through and the voltage across the load with the two batteries connected series-opposing. Reconnect the two batteries for series-opposing and run the simulation to verify your calculations.

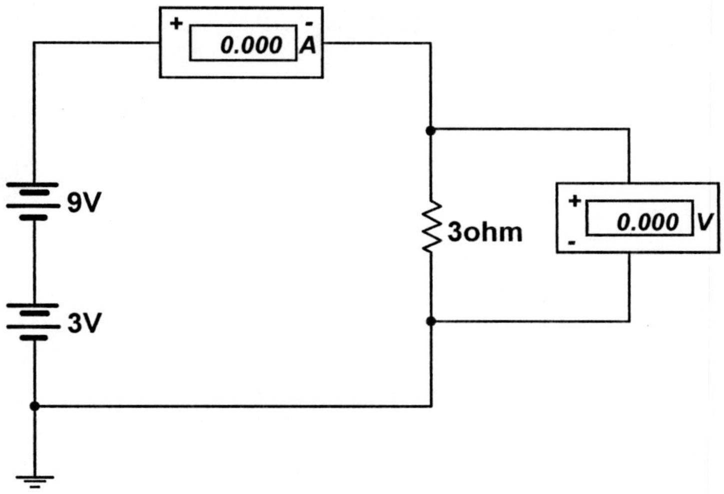

Answers

NAME ________________________________ DATE ____________________

17-8. Open the Multisim circuit file CH17 _08 shown.

(a) Turn the simulation "on" and record the no-load voltage of Battery_1 and the battery voltage and current with the load connected.

(b) Repeat with the circuit being supplied with power from Battery_2.

(c) Which battery simulates an ideal voltage source? Why?

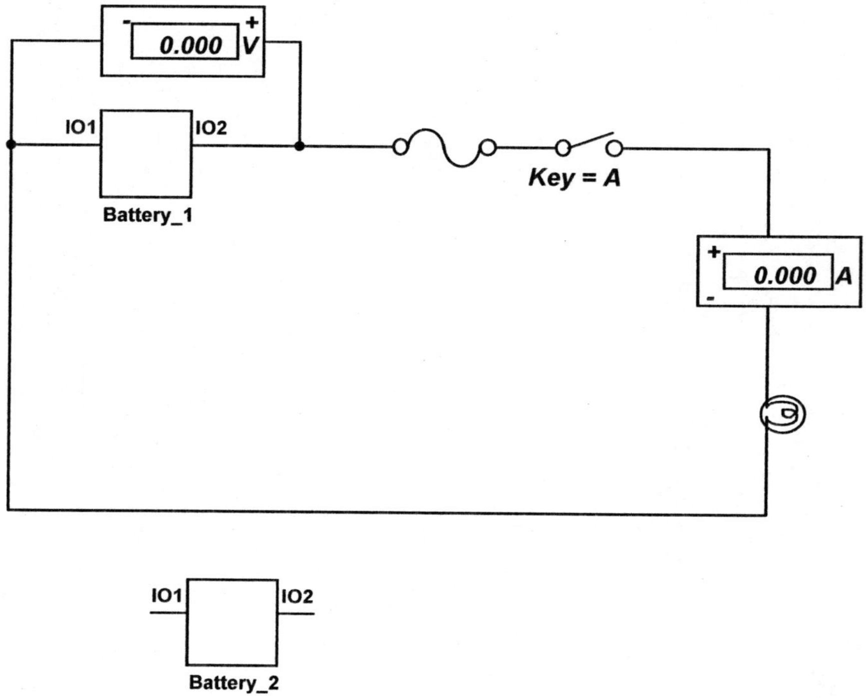

Answers

NAME ____________________________________ DATE ____________________

17-9. Open the Multisim circuit file CH17 _09 shown.
(a) Turn the simulation "on" and record the circuit's load current.
(b) Calculate the voltage drop across each of the line conductors. Measure the voltage drop across each of the line conductors to verify your calculations.
(c) Calculate the voltage across the load. Measure the load voltage to verify your calculation.

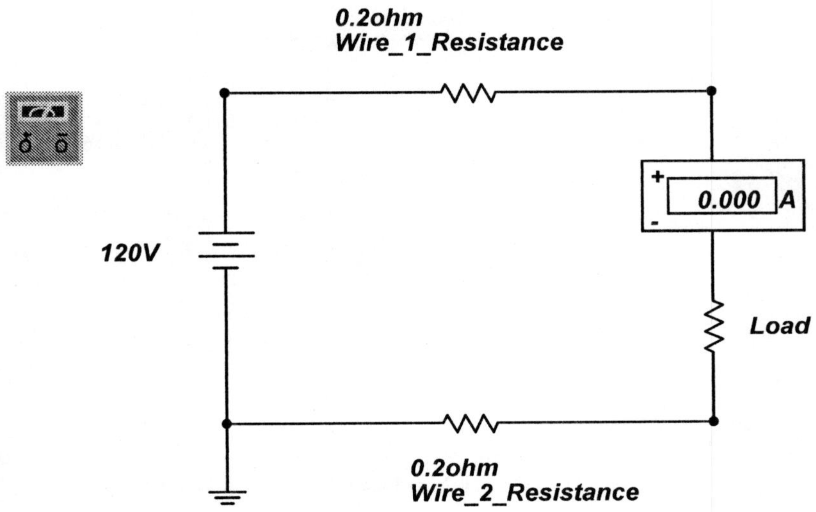

Answers

NAME ____________________ DATE ____________

17-10. Open the Multisim circuit file CH17 _10 shown.

(a) Find all the unknown values of E, I, and R, for the series circuit. Record the all given and calculated values in the table provided. Use the multimeter to measure all values of E, I, R, to verify your results

(b) Repeat for the original circuit using the defective open resistor R_2 in place of the good one.

(c) Repeat for the original circuit using the defective shorted resistor R_1 in place of the good one.

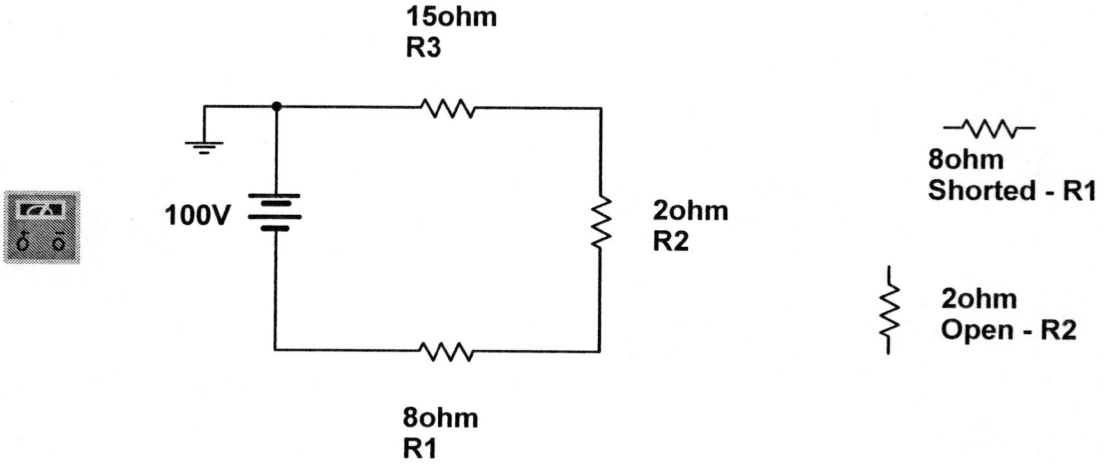

Answers (a)

	Voltage	Current	Resistance
R_1			
R_2			
R_3			
Total			

(b)

	Voltage	Current	Resistance
R_1			
R_2			
R_3			
Total			

(c)

	Voltage	Current	Resistance
R_1			
R_2			
R_3			
Total			

NAME ______________________________________ DATE ____________________

17-11. Open the Multisim circuit file CH17 _11 shown. Turn the simulation "on" and operate the circuit. Simulate an open fault in R_1 by opening the key switch. Record the change in:

(a) current flow
(b) voltage E_1
(c) voltage E_2
(d) voltage E_3
(e) total power

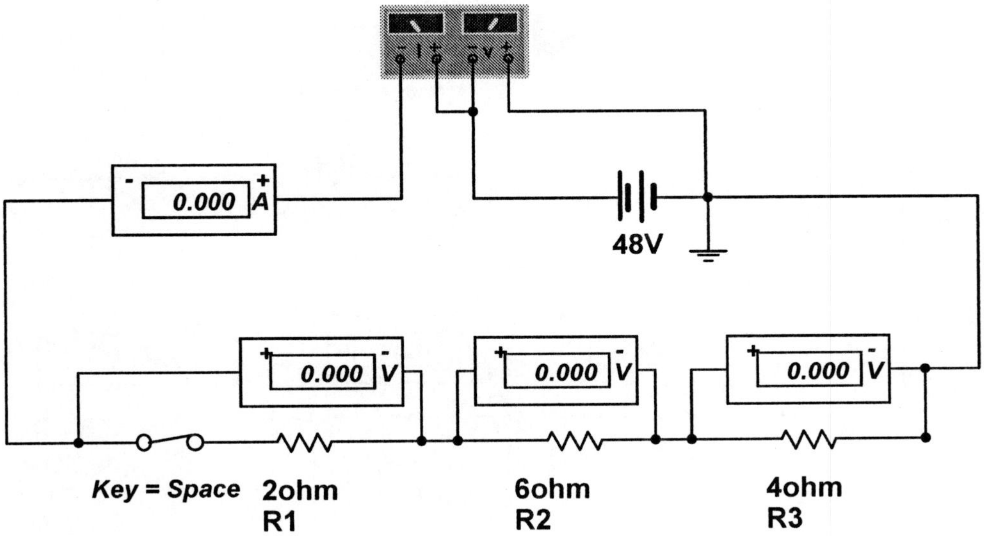

Answers

NAME ____________________ DATE ____________

17-12. Open the Multisim circuit file CH17 _12 shown. Turn the simulation "on" and operate the circuit. Simulate a shorted fault in R_2 by closing the key switch. Record the change in:

(a) current flow
(b) voltage E_1
(c) voltage E_2
(d) voltage E_3
(e) total power

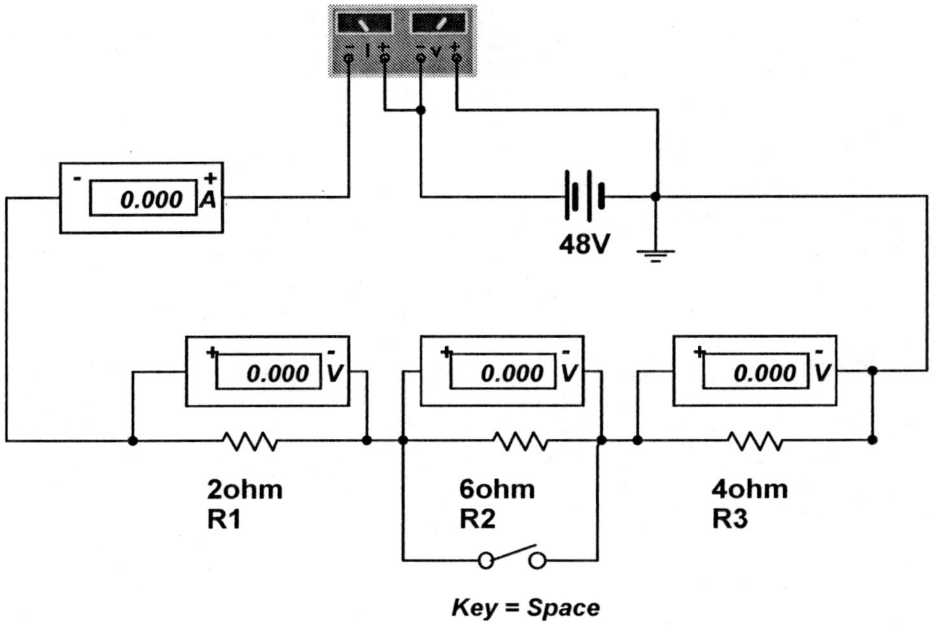

Answers

NAME ______________________________ DATE ________________

17-13. Open the Multisim circuit file CH17 _13 shown. Turn the simulation "on" and using only the voltmeter, determine which circuit is operating normal, and which circuits are faulted. Identify the faulted components and type of fault (open or short).

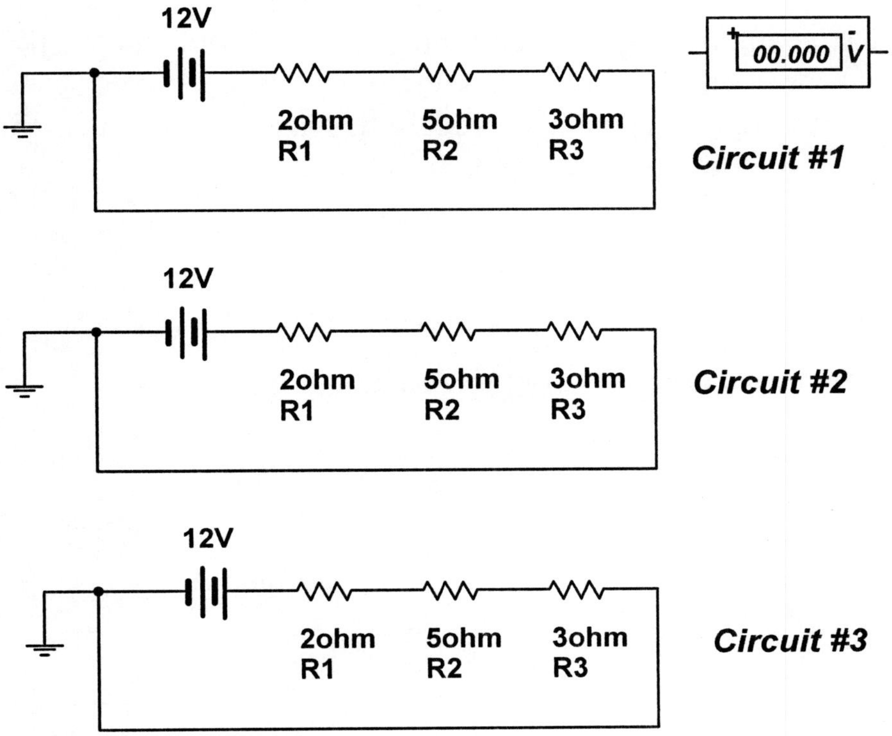

Answers

NAME ______________________ DATE ______________

17-14. Open the Multisim circuit file CH17 _14 shown. This circuit is meant to simulate a typical "signal line circuit" of an alarm system. Turn the simulation "on" and record on the following operating conditions:

(a) Sensing voltage and current with sensor switch open and open wire fault closed (no alarm condition)

(b) Sensing voltage and current with sensor switch closed and open wire fault closed (alarm condition)

(c) Sensing voltage and current with sensor switch open and open wire fault open (fault condition)

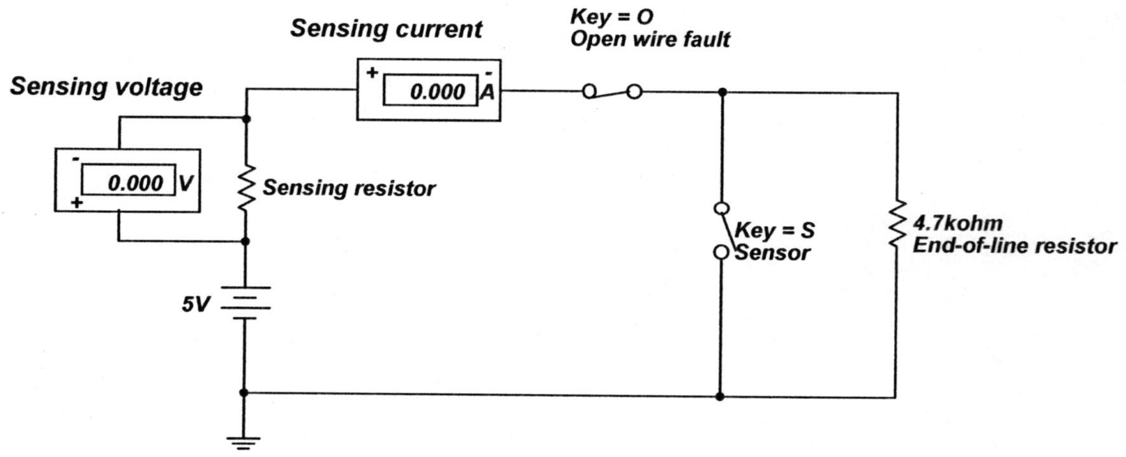

Answers

NAME ______________________ DATE ______________

17-15 Open the Multisim circuit file CH17 _15 shown. This circuit illustrates how the series circuit is used to heat or warm an oven. The circuit consists of a heating element, warming element, ON/OFF switch, and a temperature switch. The temperature switch is closed. until its preset temperature is reached, at which time it opens.

(a) Calculate the value of the heating and warming current. Use the multimeter to measure the value of the heating and warming current.

(b) Calculate the value of the heating and warming power. Use the wattmeter to measure the value of the heating and warming power.

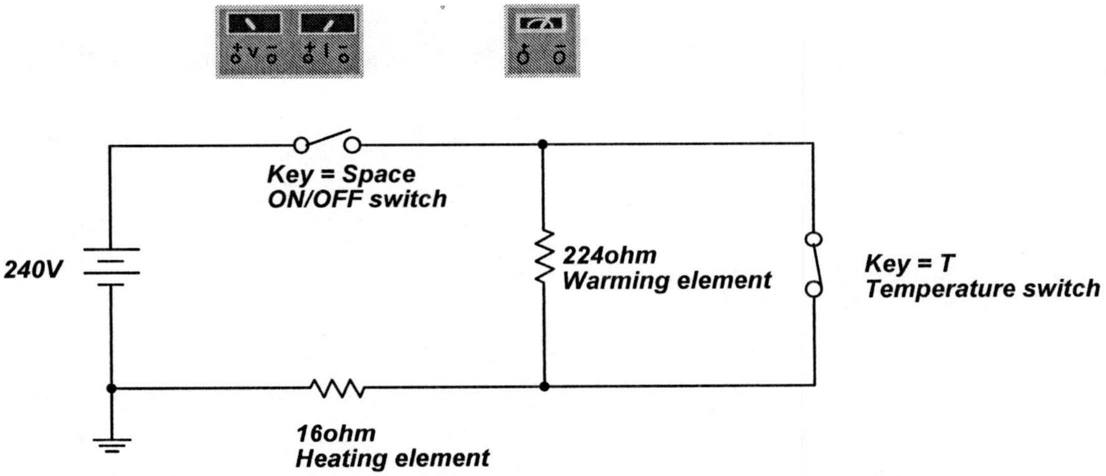

Answers

NAME ______________________________ DATE ________________

UNIT 18

SOLVING THE DC PARALLEL CIRCUIT

Multisim Simulation Assignments

18-1. Open the Multisim circuit file CH18 _01 shown. Use the multimeter for all voltage, current, and resistance measurements. Use the wattmeter for all power measurements.

(a) Measure the source voltage and the voltage across each resistor.

(b) Based on your readings, state the voltage characteristics of a parallel circuit.

(c) Measure the source current and the current flow through each resistor.

(d) Based on your current readings state the current characteristics of a parallel circuit.

(e) Measure the total circuit resistance and the resistance of each resistor.

(f) Based on your resistance readings state the resistance characteristics of a parallel circuit.

(g) Measure the power dissipation of each resistor and that of the total circuit.

(h) Based on your wattmeter readings state the power characteristics of a parallel circuit.

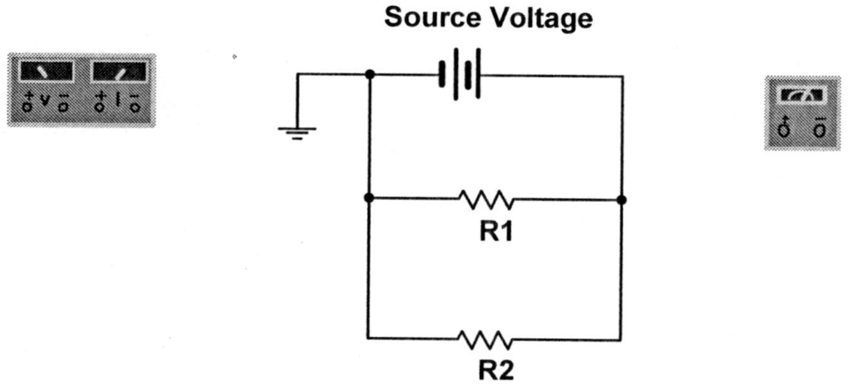

Answers

NAME ______________________________ DATE ________________

18-2. Open the Multisim circuit file CH18 _02 shown. Use the multimeter for all voltage, current, and resistance measurements.

(a) Calculate the value of the total resistance. Measure the total resistance to verify your answer.

(b) Calculate the value of the total current flow and the current flow through each of the resistors. Measure these current flow values to verify your answers.

(c) What is the value of the voltage across R_1 and R_2? Measure the these voltage drops to verify your answer.

(d) Why is the current flow through R_1 the same as that through R_2?

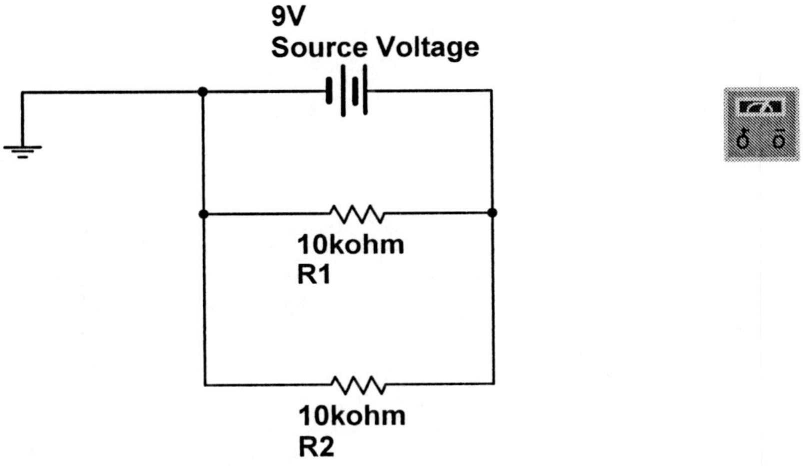

Answers

NAME ____________________ DATE ____________

18-3. Open the Multisim circuit file CH18 _03 shown. Use the multimeter for all voltage, current, and resistance measurements.

(b) Calculate the value of the total resistance. Measure the total resistance to verify your answer.

(b) Calculate the value of the total current flow and the current flow through each of the resistors. Measure these current flow values to verify your answers.

(c) What is the value of the voltage across R_1 and R_2? Measure these voltage drops to verify your answer.

(d) Why is the current flow through R_1 greater that through R_2?

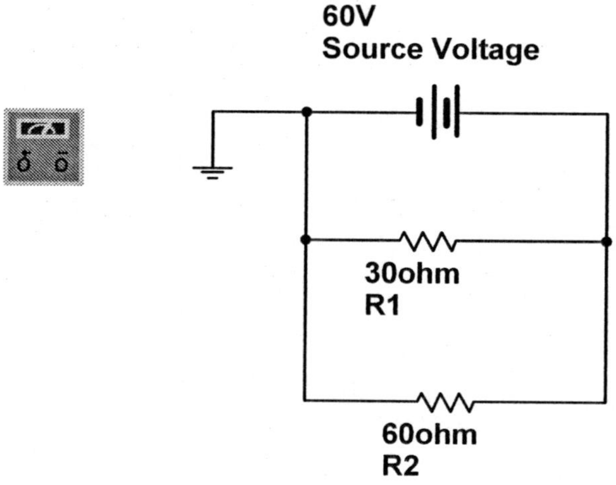

Answers

NAME ______________________________ DATE ________________

18-4. (a) Open the Multisim circuit file CH18 _04 shown. Find all the unknown values of E, I, R, and P, for the parallel circuit. Record the all given and calculated values in the table provided.

(b) Use the multimeter and wattmeter to measure all values of E, I, R, P, and record your readings in the table provided.

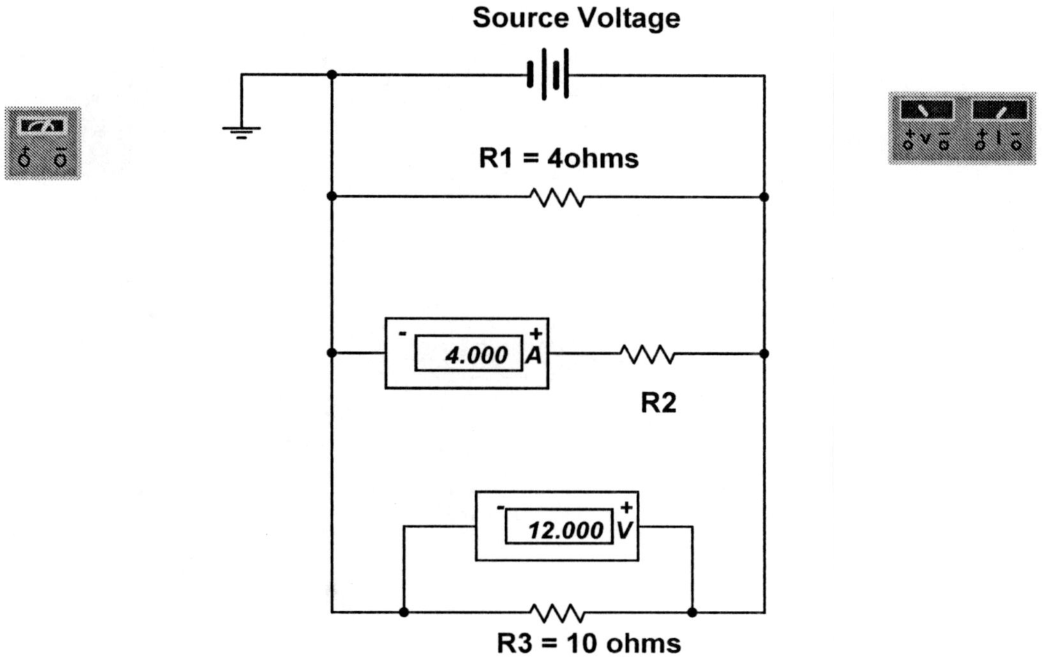

Answers

(a)

	Voltage	Current	Resistance	Power
R_1				
R_2				
R_3				
Total				

(b)

	Voltage	Current	Resistance	Power
R_1				
R_2				
R_3				
Total				

NAME ______________________________ DATE ________________

18-5. (a) Open the Multisim circuit file CH18 _05 shown. Find all the unknown values of E, I, R, and P, for the parallel circuit. Record the all given and calculated values in the table provided.

(b) Use the multimeter and wattmeter to measure all values of E, I, R, P, and record your readings in the table provided.

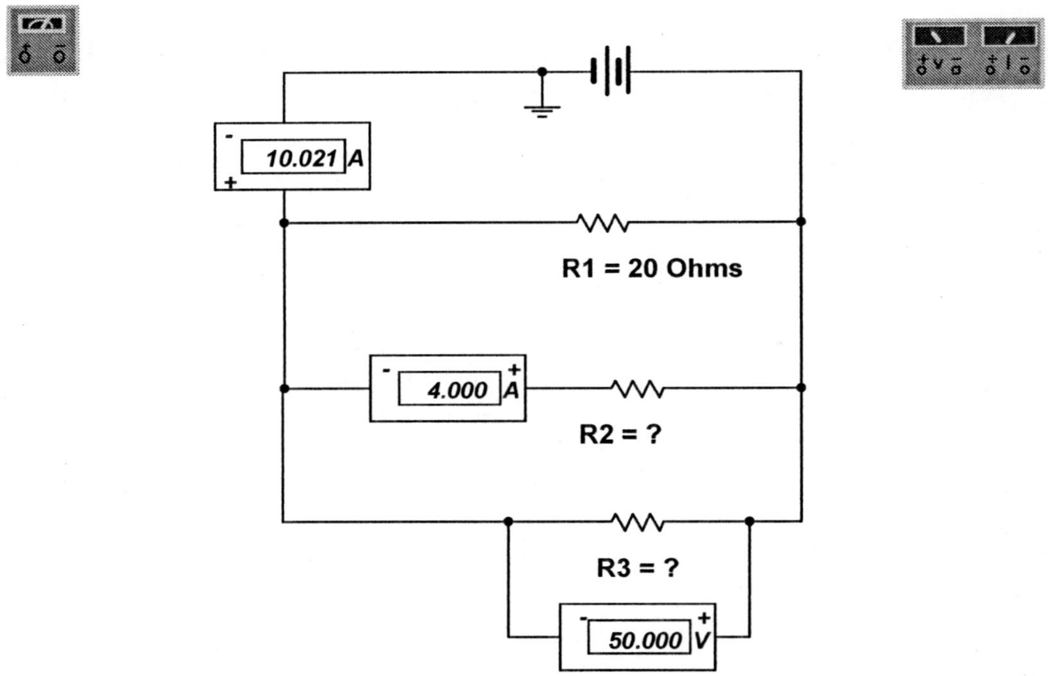

Answers

(a)

	Voltage	Current	Resistance	Power
R_1				
R_2				
R_3				
Total				

(b)

	Voltage	Current	Resistance	Power
R_1				
R_2				
R_3				
Total				

NAME ______________________________ DATE ________________

18-6. Open the Multisim circuit file CH18 _06 shown. What is the polarity of point "A" relative to point "B"? Use the multimeter to verify your answer.

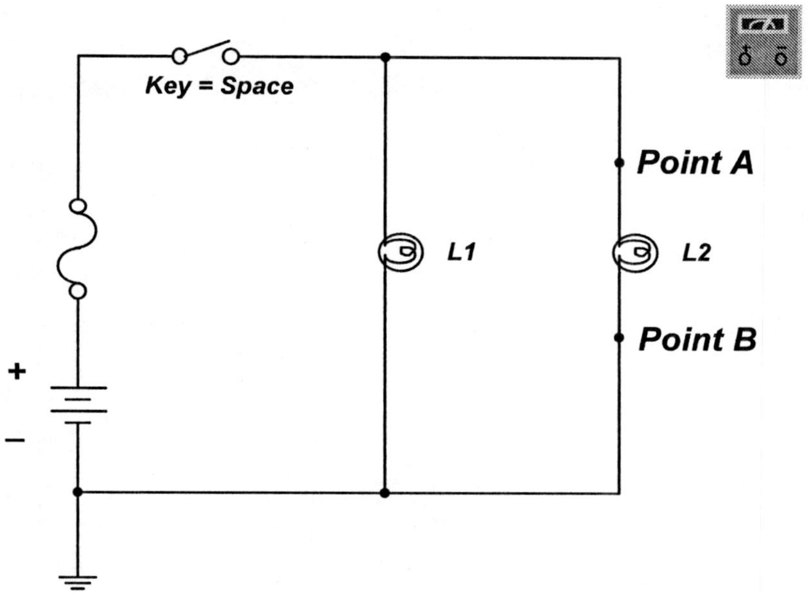

Answer

NAME ________________________________ DATE ________________

18-7. Open the Multisim circuit file CH18 _07 shown.

(a) Find all the unknown values of E, I, and R, for the parallel circuit. Record the all given and calculated values in the table provided. Use the multimeter to measure all values of E, I, R, to verify your results.

(b) Repeat for the original circuit using the defective open resistor R_1 in place of the good one.

(c) Operate the circuit with the defective shorted resistor R_1 in place of the good one. What happens? Why?

(d) Record the total resistance of the circuit with the defective shorted resistor R_1 connected.

Answers

(a)

	Voltage	**Current**	**Resistance**
R_1			
R_2			
R_3			
Total			

(b)

	Voltage	**Current**	**Resistance**
R_1			
R_2			
R_3			
Total			

NAME ______________________________________ DATE ____________________

18-8 Open the Multisim circuit file CH18 _08 shown.
Turn the simulation "on" and operate the circuit. Simulate an open fault in R_1 by opening the key switch. Record the change in:

(a) I_T
(b) I_1
(c) I_2
(d) E_T
(e) E_1
(f) E_2
(g) P_T

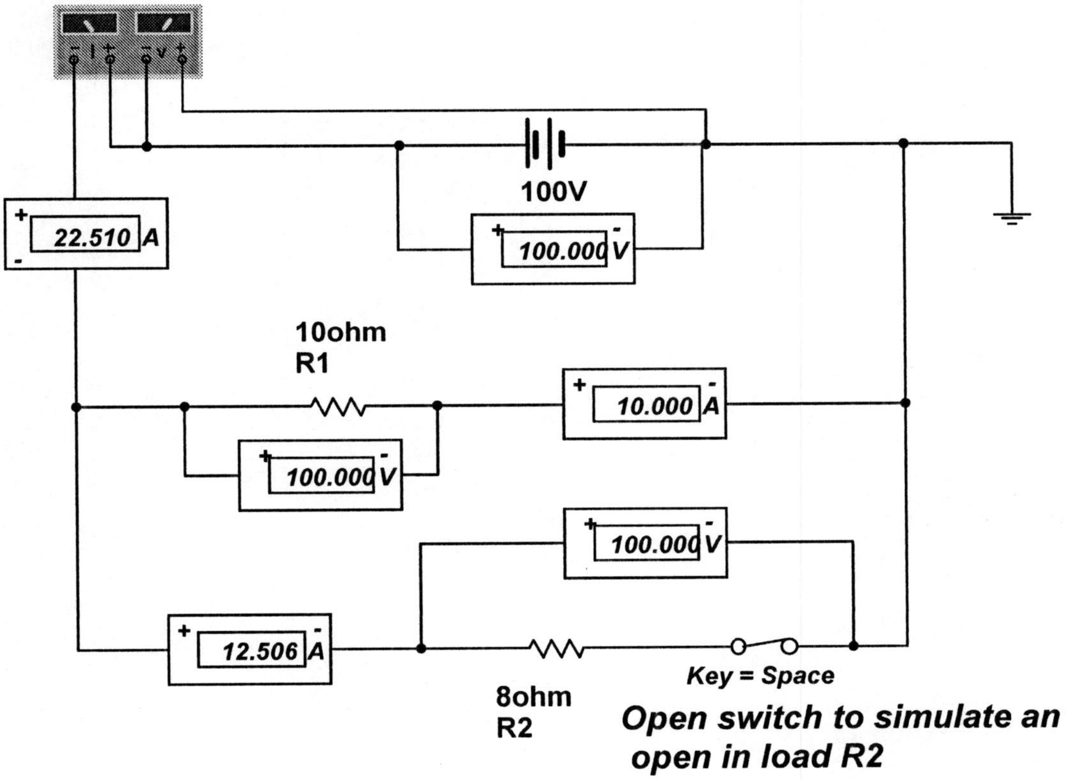

Answers

NAME ____________________ DATE ____________

18-9 Open the Multisim circuit file CH18 _09 shown. Using the multimeter determine which circuit is operating normally, and which circuits are faulty. Identify the faulted components and type of fault (open or short).

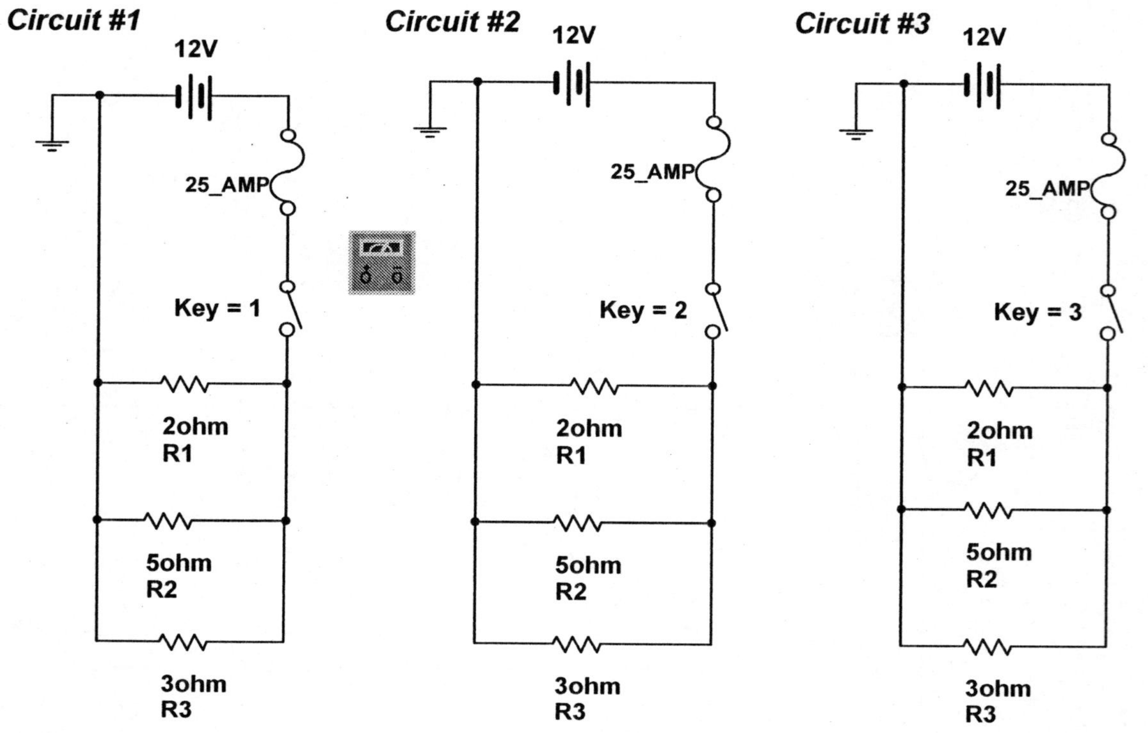

Answers

NAME ______________________________ DATE ______________

18-10 Open the Multisim circuit file CH18 _10 shown.

(a) Calculate how many 15-Ω resistive loads can be connected in parallel before the maximum current rating of the fuse is exceeded.

(b) Record the total current flow with the maximum number of 15-Ω loads connected in parallel.

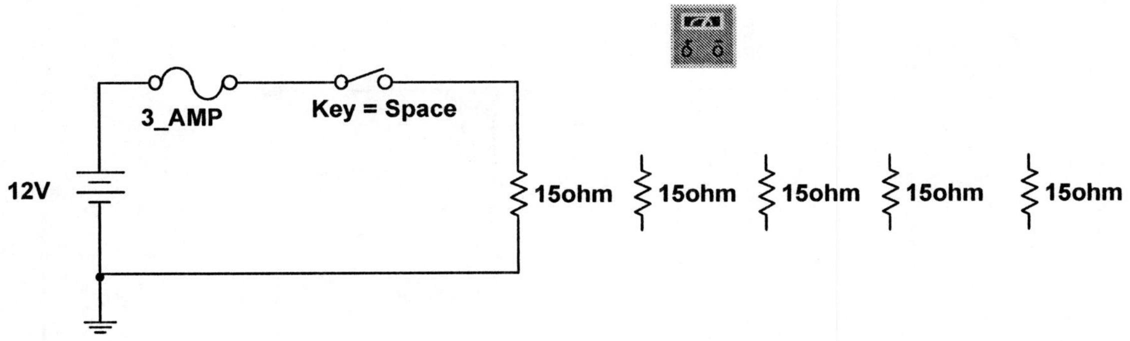

Answer

NAME ________________________________ DATE ________________

18-11 (a) Open the Multisim circuit file CH18 _11 shown. Calculate the value of the current flow through each of the lamps and that of the line.

(b) Measure the value of the current flow through each of the lamps and that of the line.

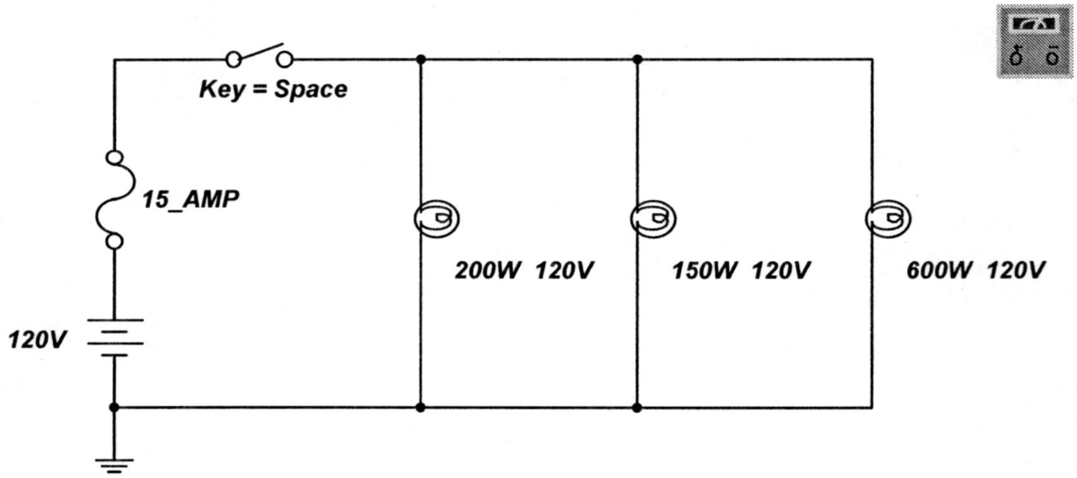

Answers

NAME ______________________________ DATE ________________

18-12 Open the Multisim circuit file CH18 _12 shown.

Determine the total circuit resistance as seen by the power supply for each of the following switching scenarios:

(a) S_1 open, S_2 closed, and S_3 closed.

(b) S_1 closed, S_2 open, and S_3 open.

(c) S_1 closed, S_2 closed, and S_3 open.

(d) S_1 closed, S_2 open, and S_3 closed.

(e) S_1 closed, S_2 closed, and S_3 closed.

(Measure the actual resistance values to verify your results)

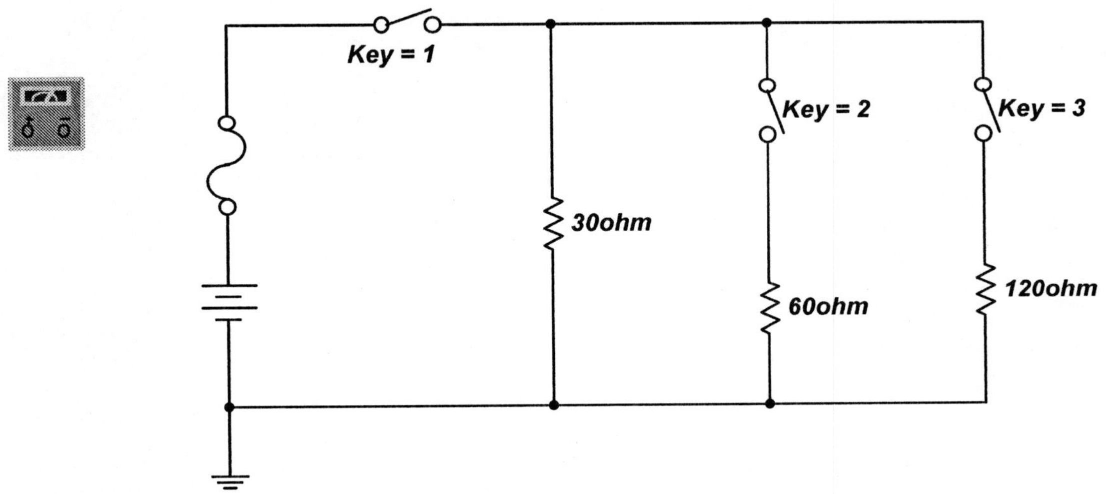

Answers

NAME ______________________________ DATE ____________________

18-13 Open the Multisim circuit file CH18 _13 shown.
This circuit consists of a 10-V battery connected to ten equal-value resistors connected in parallel. If the total power dissipated by the circuit is 40 watts determine the value of each resistor.

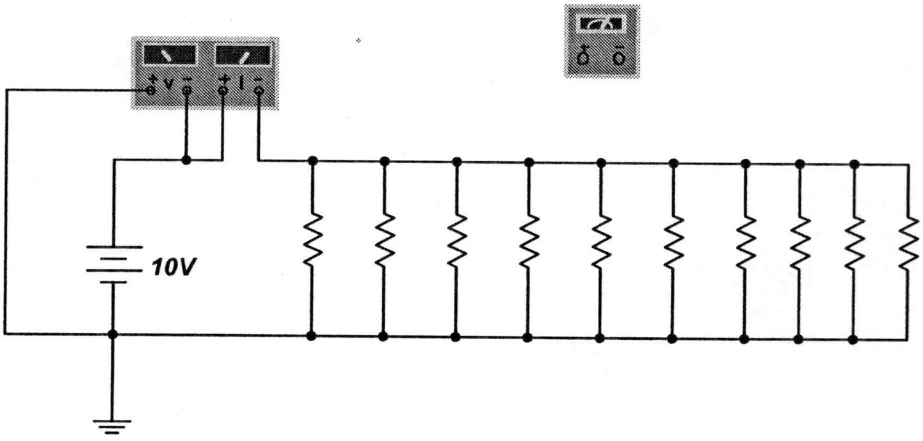

Answer

NAME ______________________________ DATE ________________

UNIT 19

SOLVING THE DC SERIES-PARALLEL CIRCUIT

Multisim Simulation Assignments

19-1. Open the Multisim circuit file CH19 _01 shown. Find all the unknown values of E, I, and R, for the series-parallel circuit. Record all the given and calculated values in the table provided. Use the multimeter to measure all values of E, I, and R, to verify your results.

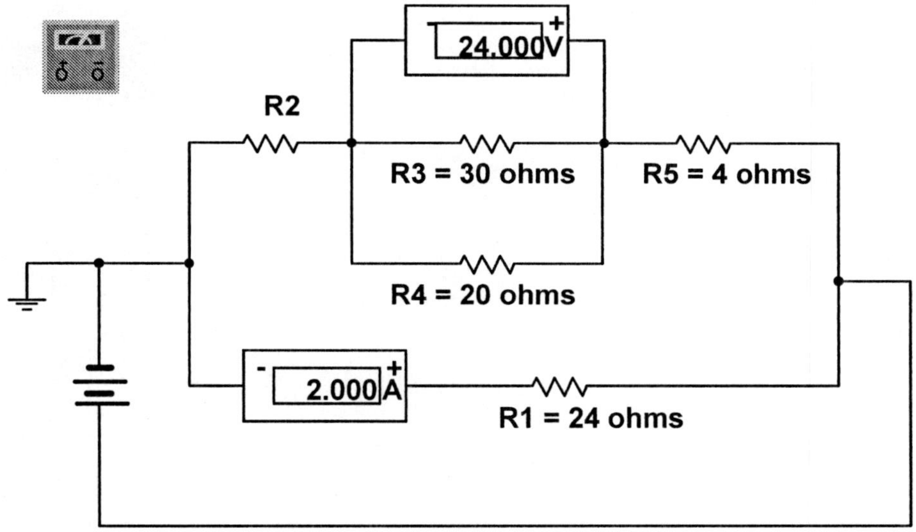

<u>Answers</u>

	Voltage	Current	Resistance
R_1			
R_2			
R_3			
R_4			
R_5			
Total			

NAME ______________________________ DATE ________________

19-2. Open the Multisim circuit file CH19 _02 shown. Determine, using Kirchhoff's current and voltage laws, I_1, E_2, I_4, and E_4. Reconnect the voltmeters and ammeters to verify your answers.

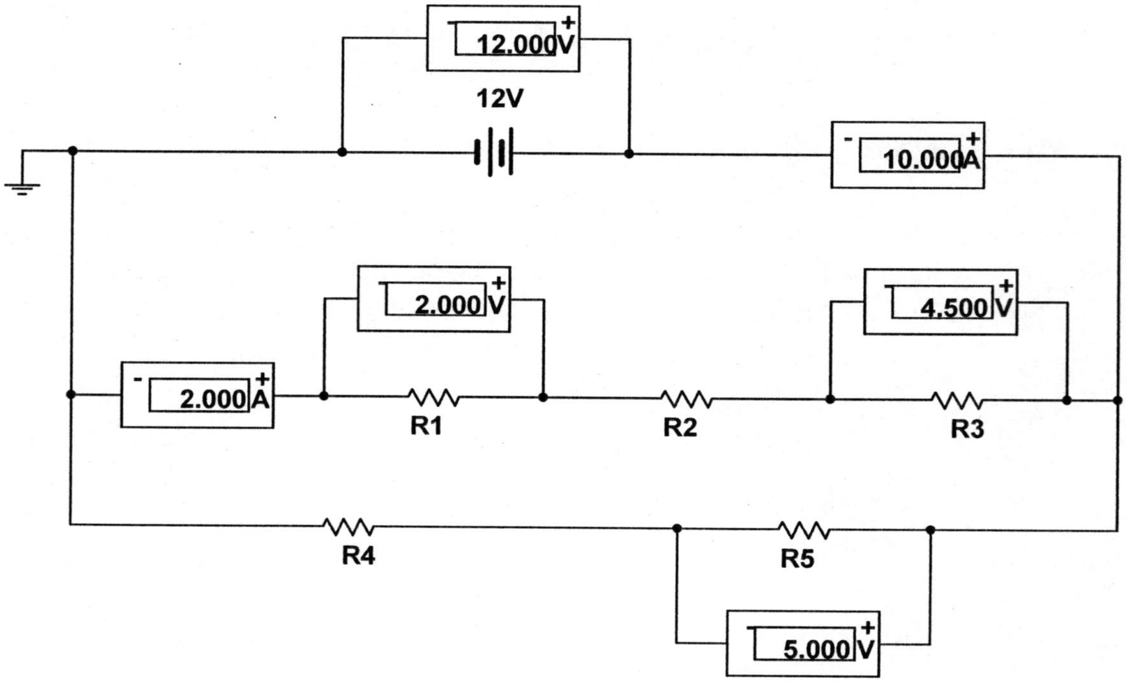

Answers

NAME ______________________________ DATE ________________

19-3. Open the Multisim circuit file CH19 _03 shown. Find all the unknown values of E, I, R, and P, for the circuit. Record all the given and calculated values in the table provided. Connect the multimeter and wattmeter to verify your results.

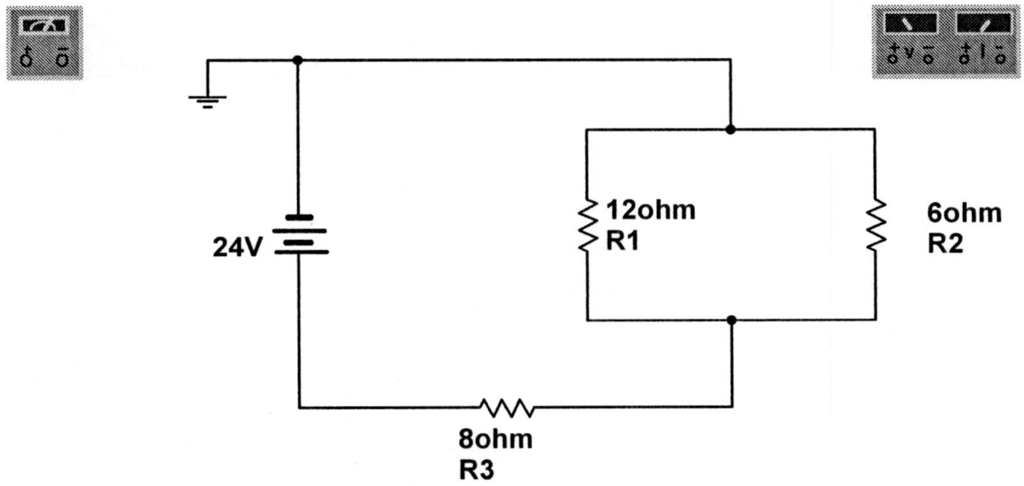

Answers

	Voltage	Current	Resistance	Power
R_1				
R_2				
R_3				
Total				

NAME ______________________________ DATE ________________

19-4. Open the Multisim circuit file CH19 _04 shown.

a) Find all the unknown values of E, I, and R, for the circuit. Record the all given and calculated values in the table provided. Use the multimeter to measure all values of E, I, and R, to verify your results.

(b) Repeat for the original circuit using the defective open resistor R_3 in place of the good one.

(c) Repeat for the original circuit using the defective shorted resistor R_1 in place of the good one.

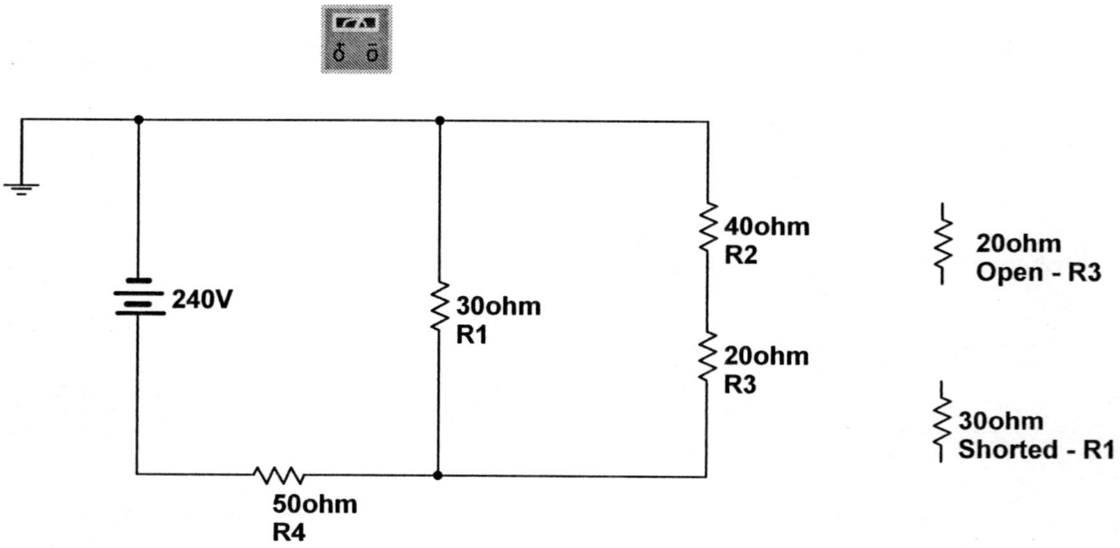

<u>Answers</u> (a) Normal circuit

	Voltage	Current	Resistance
R_1			
R_2			
R_3			
R_4			
Total			

NAME ____________________ DATE ____________

(b) Opened R_3 resistor

	Voltage	Current	Resistance
R_1			
R_2			
R_3			
R_4			
Total			

(b) Shorted R_1 resistor

	Voltage	Current	Resistance
R_1			
R_2			
R_3			
R_4			
Total			

NAME ____________________ DATE ____________

19-5. Open the Multisim circuit file CH19 _05 shown.

(a) Use the multimeter to record the value of the voltage drop across each of resistor load.

(b) Based on your voltage readings, which resistor is at fault. Given a reason for your answer and classify the fault condition as an open or short.

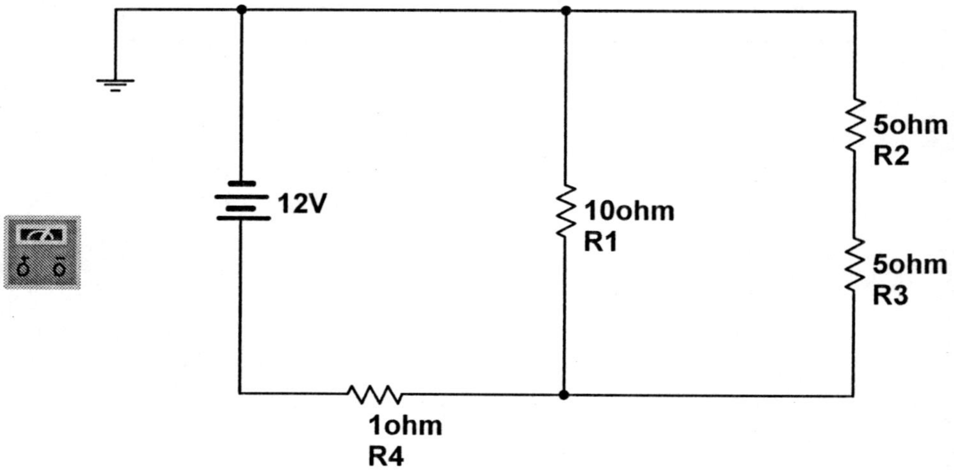

Answers

NAME ________________________________ DATE ____________________

19-6. Open the Multisim circuit file CH19 _06 shown.

(a) Use the multimeter to record the value of the voltage drop across each of resistor load.

(b) Based on your voltage readings, which resistor is at fault. Given a reason for your answer and classify the fault condition as an open or short.

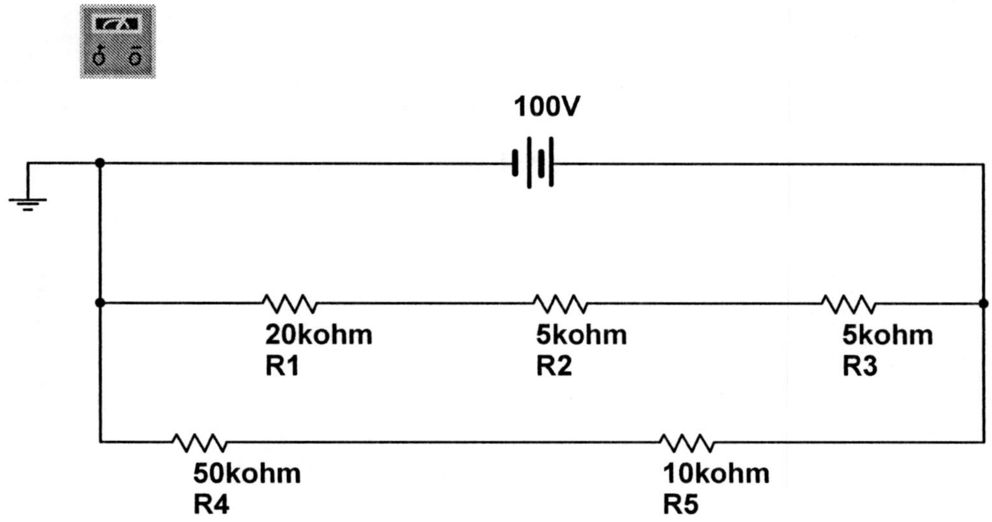

Answers

NAME ______________________________ DATE ________________

19-7. Open the Multisim circuit file CH19 _07 shown.
Find the value of the power loss in the feed line and return line of the distribution circuit shown. Connect the wattmeter to verify your results.

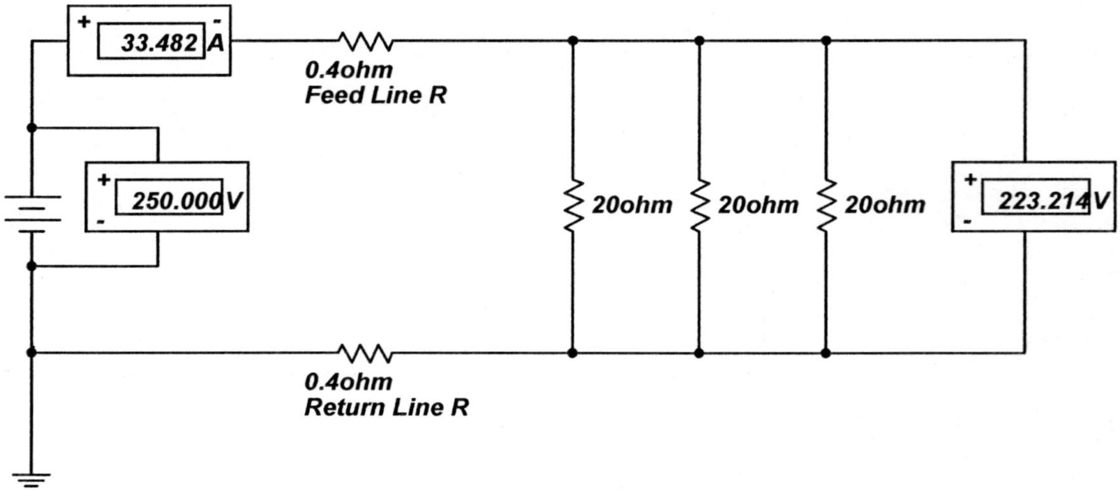

Answers

NAME ______________________________ DATE __________________

UNIT 20

NETWORK THEOREMS

Multisim Simulation Assignments

20-1. (a) Open the Multisim circuit file CH20 _01 shown. Calculate the output voltage of the real voltage source with the R_L 5Ω load connected. Turn the simulation on and measure the output voltage to verify your calculation.
(b) Repeat for the 10Ω load resistor.
(c) Repeat for the 2Ω load resistor.
(d) What happens to output voltage (greater or less output) when the load on the circuit is increased and decreased?

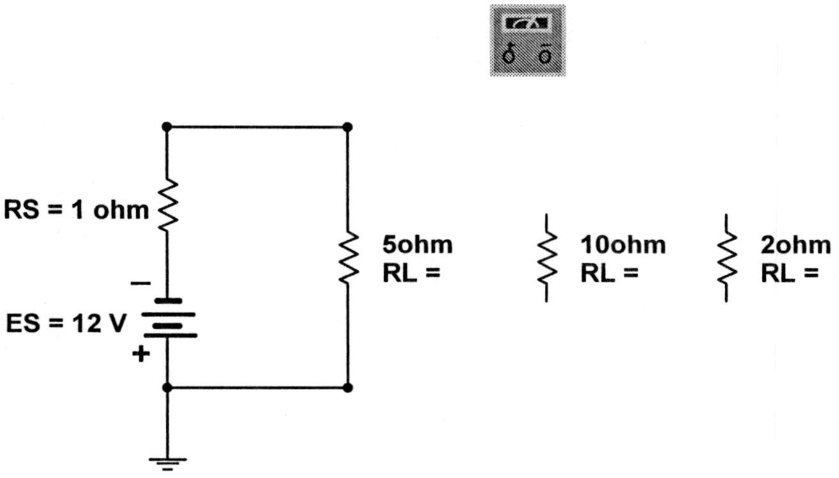

Answers

NAME ______________________________ DATE ________________

20-2. (a) Open the Multisim circuit file CH20 _02 shown. Calculate the amount of current flow through R_L produced by the current source, with the 5Ω load connected. Turn the simulation on and measure the current flow through R_L to verify your calculation.
(b) Repeat for the 10Ω load resistor.
(c) Repeat for the 2Ω load resistor.

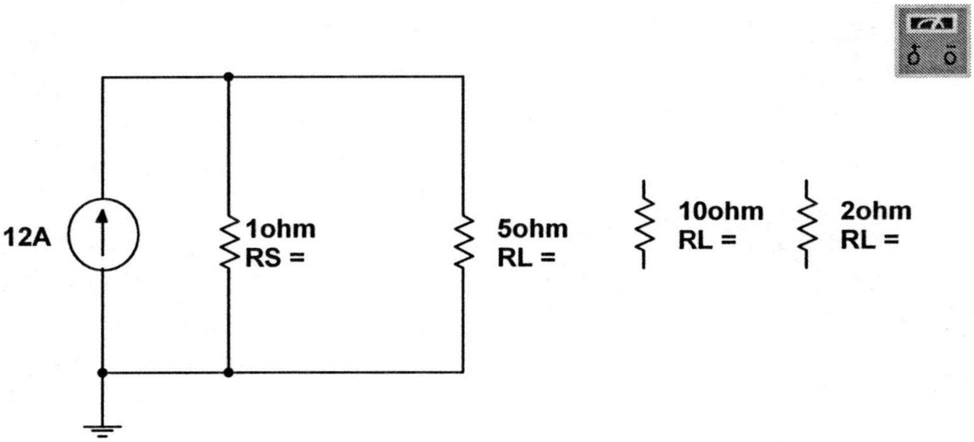

Answers

NAME ______________________________ DATE ________________

20-3. (a) Open the Multisim circuit file CH20 _03 shown. Calculate the amount of current flow through and voltage drop across R_1 and R_2 using the superposition theorem.

(b) Turn the simulation on and measure the current and voltage of R_1 and R_2.

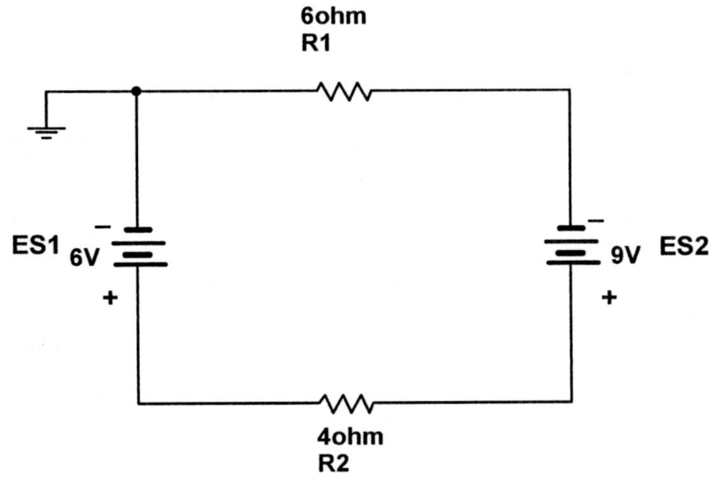

Answers

NAME ______________________________ DATE ________________

20-4. (a) Open the Multisim circuit file CH20 _04 shown. Calculate the amount of current flow through R_3 using the superposition theorem.

(b) Turn the simulation on and measure the current of R_3.

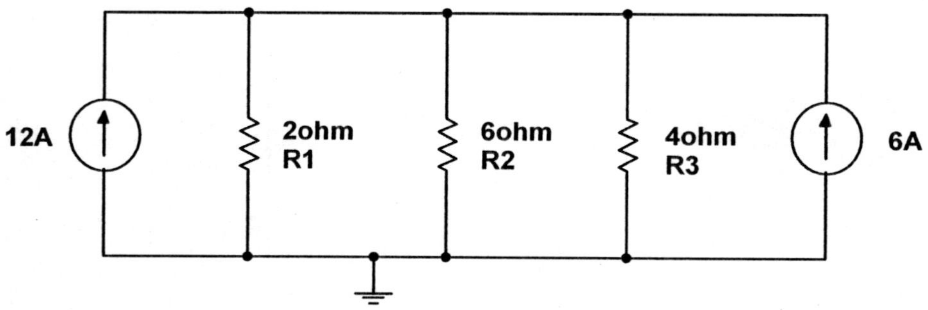

Answers

NAME ___________________________________ DATE __________________

20-5. (a) Open the Multisim circuit file CH20 _05 shown. Calculate the amount of current I_1, I_2, and I_3, for the circuit using the superposition theorem.

(b) Turn the simulation on and measure I_1, I_2, and I_3.

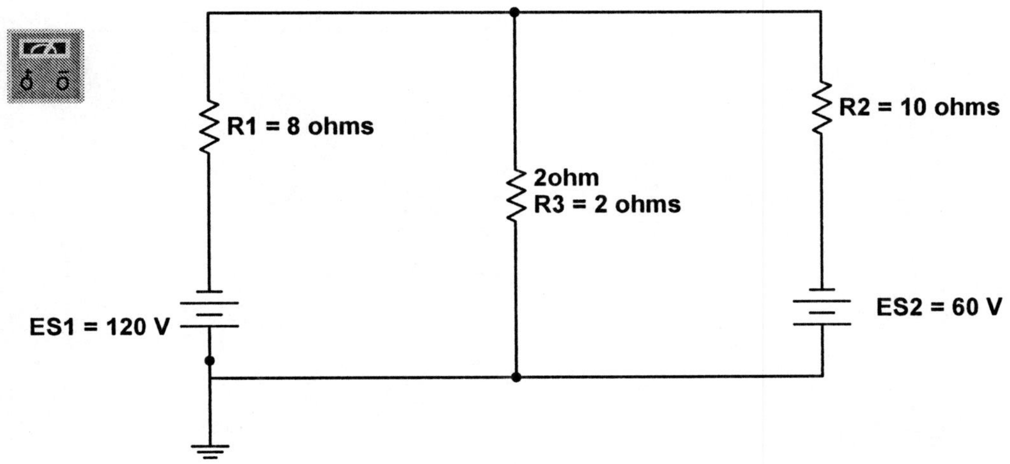

Answers

NAME ________________________________ DATE ________________

20-6. (a) Open the Multisim circuit file CH20 _06 shown. Calculate the amount of current flow through R_2 using the superposition theorem.

(b) Turn the simulation on and measure the current of R_2.

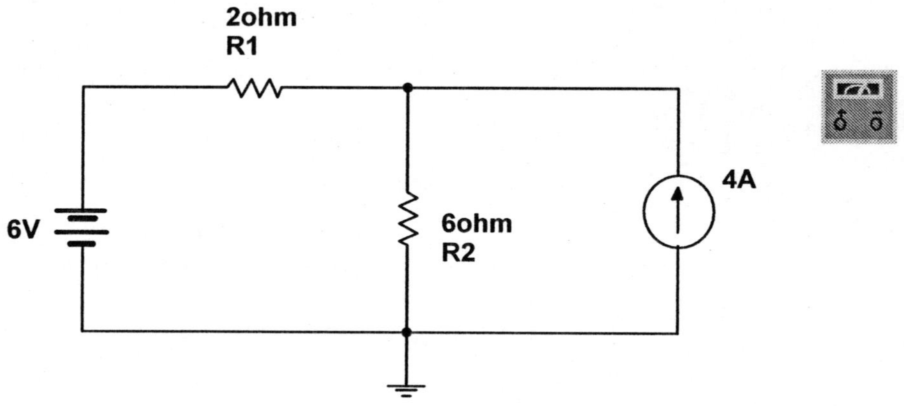

Answers

NAME ______________________________ DATE __________________

20-7. (a) Open the Multisim circuit file CH20 _07 shown. Calculate the amount of current flow through the common wire using the superposition theorem. Turn the simulation on and measure the current flow through the common wire to verify your calculation.

(b) Repeat for an R_2 value of 20Ω.

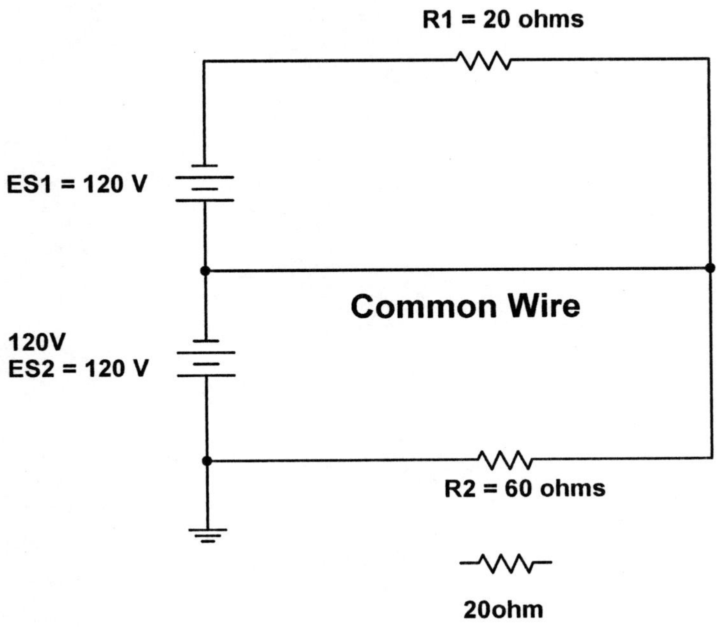

Answers

NAME ______________________________ DATE ________________

20-8. (a) Open the Multisim circuit file CH20 _08 shown. Calculate the current flow through and the voltage across R_L using Thevenin's theorem. Turn the simulation on and measure the current and voltage of R_L to verify your answer.

(b) Repeat for a load resistor of 9Ω.

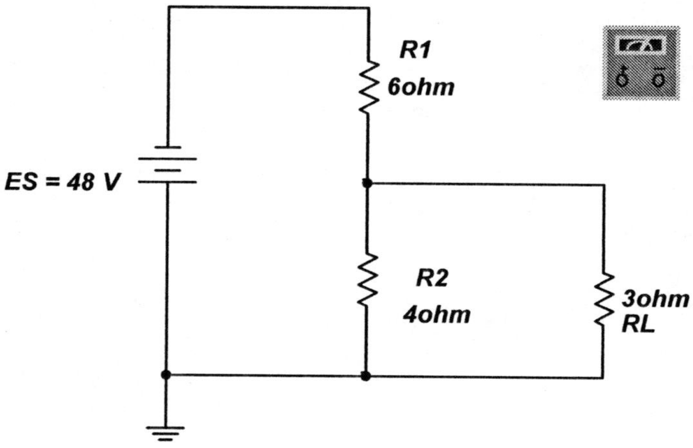

Answers

NAME ______________________________ DATE ______________

20-9. (a) Open the Multisim circuit file CH20 _09 shown. Use Thevenin's theorem to determine the current through and the voltage across R_L, of the dual-voltage circuit shown.

(b) Turn the simulation on and measure the current through and the voltage across R_L.

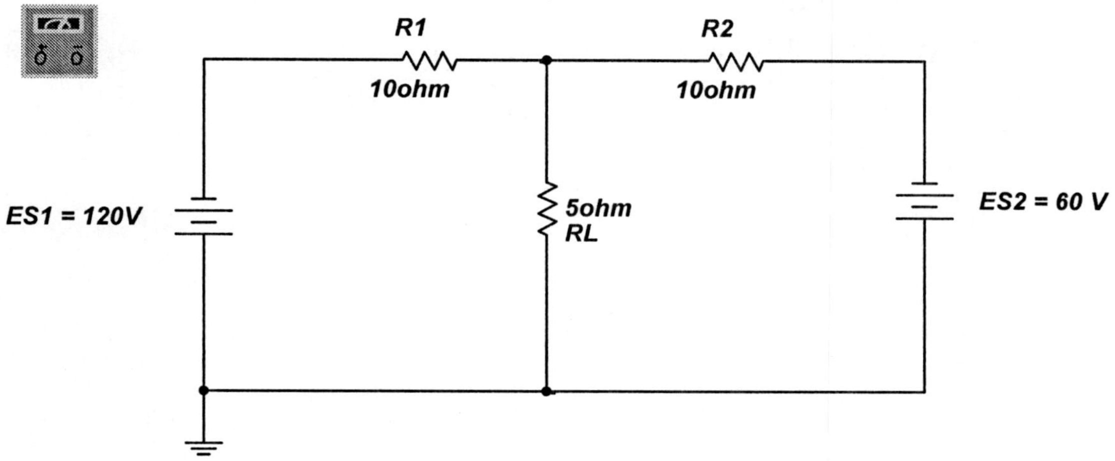

Answers

NAME ________________________________ DATE ________________

20-10. (a) Open the Multisim circuit file CH20 _10 shown. Use the multimeter to measure the Thevenin resistance (R_{TH}) and the Thevenin voltage (E_{TH}) for the circuit.

(b) Use the Thevenin equivalent circuit to determine the current through and the voltage across R_L. Turn the simulation on and measure the current through and the voltage across R_L to verify your answer.

(c) Repeat for a R_L of 5Ω.

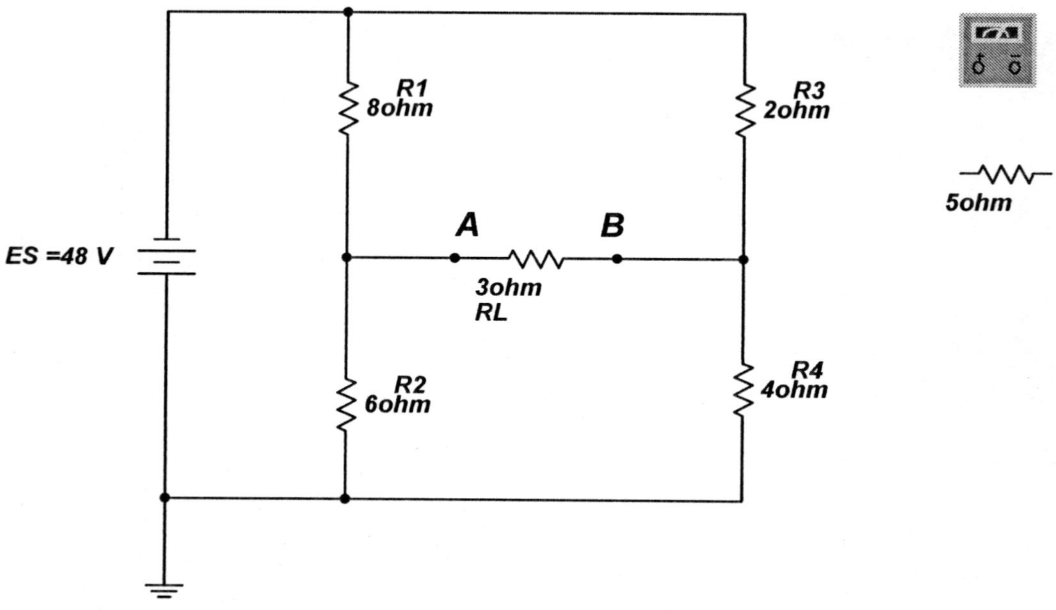

Answers

NAME ______________________________ DATE ________________

20-11. (a) Open the Multisim circuit file CH20 _11 shown. Derive the Norton equivalent circuit for the circuit shown. Calculate the load current and voltage for the 12Ω load resistor R_L. Turn the simulation on and measure the current through and the voltage across R_L to verify your answer.
(b) Repeat for a 8Ω load resistor.

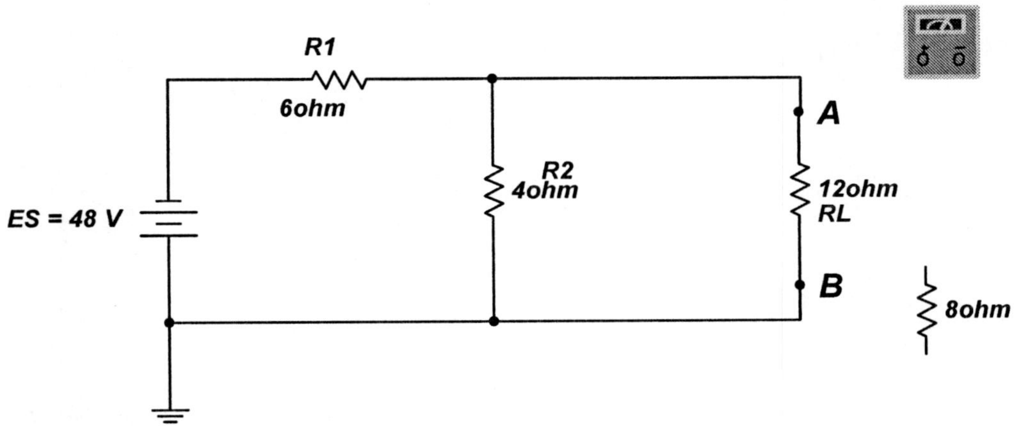

Answers

NAME ______________________ DATE ______________

20-12. (a) Open the Multisim circuit file CH20 _12 shown. Derive the Norton equivalent circuit for the circuit shown. Calculate the load current and voltage for the 5Ω load resistor R_L. Turn the simulation on and measure the current through and the voltage across R_L to verify your answer.

(c) Derive the Thevenin equivalent for the circuit.

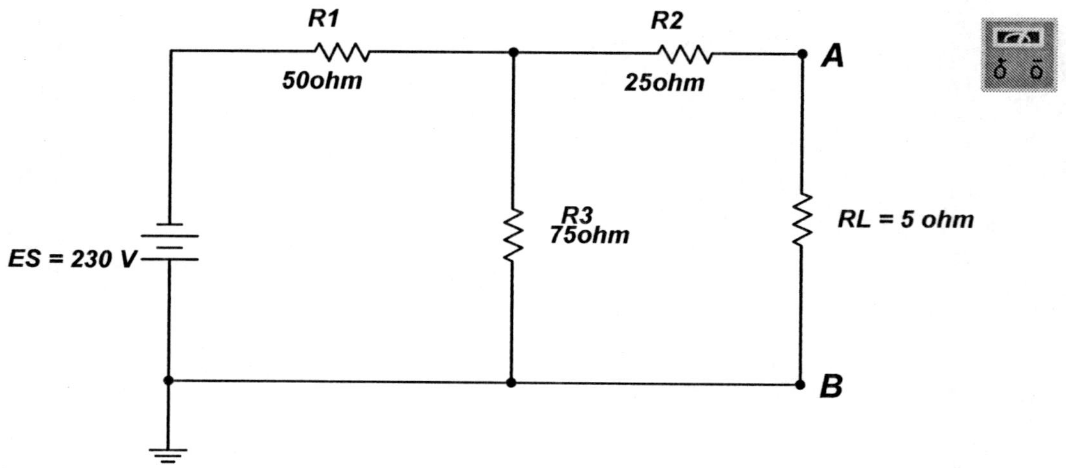

Answers

NAME ____ DATE ____________________

UNIT 21

DIRECT CURRENT AND ALTERNATING CURRENT

Multisim Simulation Assignments

21-1. Open the Multisim circuit file CH21 _01 shown. Turn the simulation on and double click the oscilloscope icon to view the scope display and front panel control screen. According to the timebase setting what is the value of the period for the AC waveform displayed?

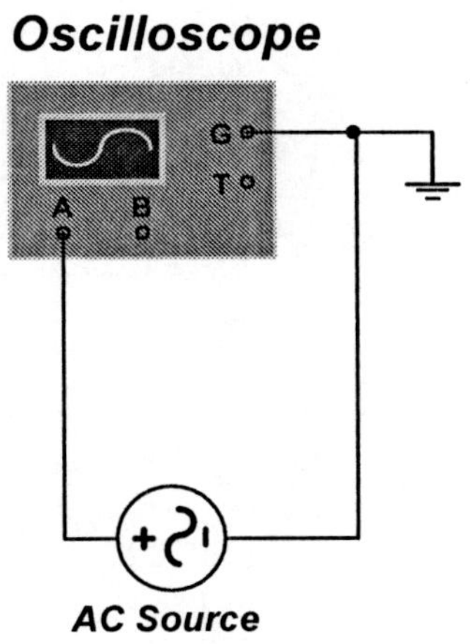

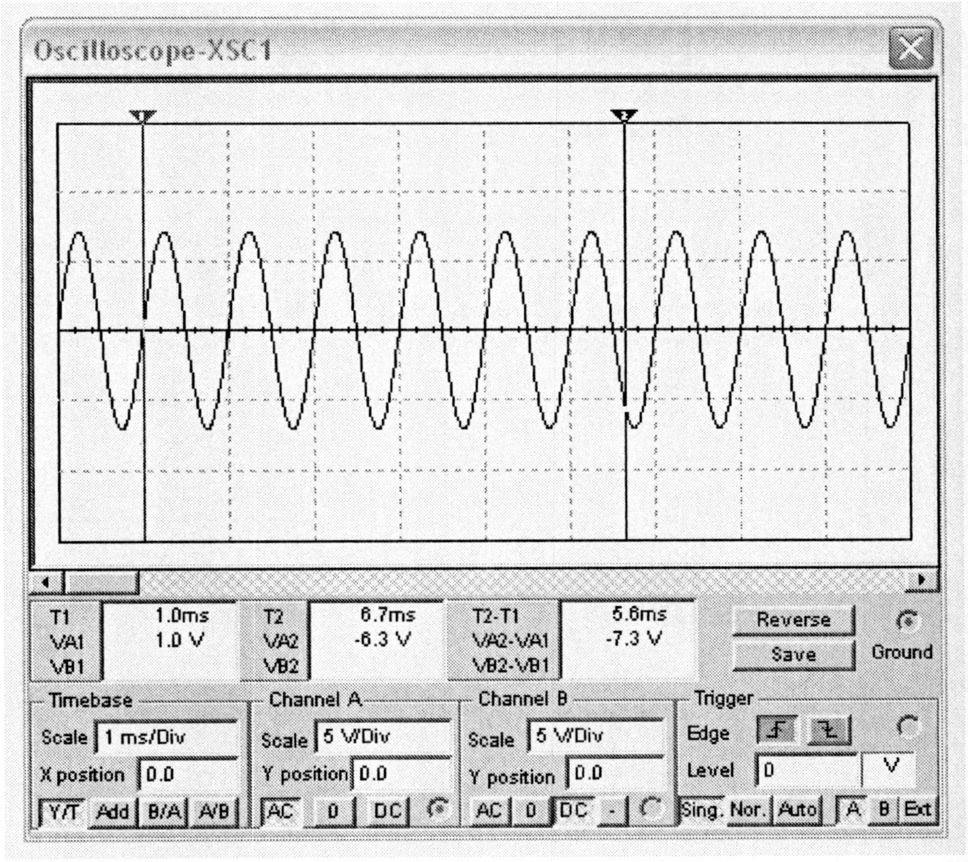

Answer

NAME ____________________________________ DATE ________________

21-2. Open the Multisim circuit file CH21 _02 shown. Turn the simulation on and double click the oscilloscope icon to view the scope display and front panel control screen. According to the 5V/D setting of Channel "A" what is the peak value of the AC waveform displayed? Verify your answer by double clicking on the AC source icon to view the voltage amplitude setting of the waveform.

Oscilloscope

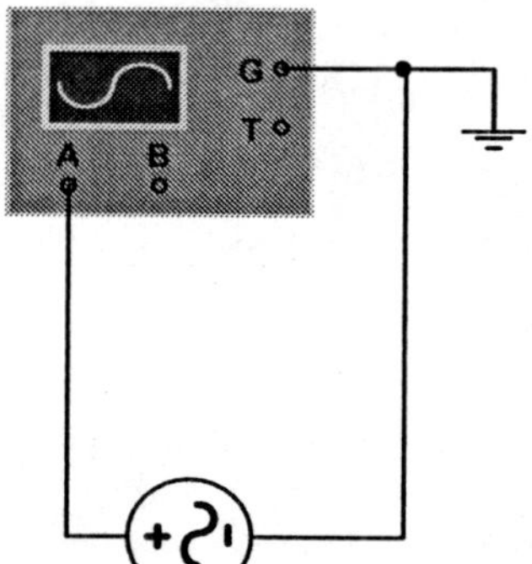

AC Source

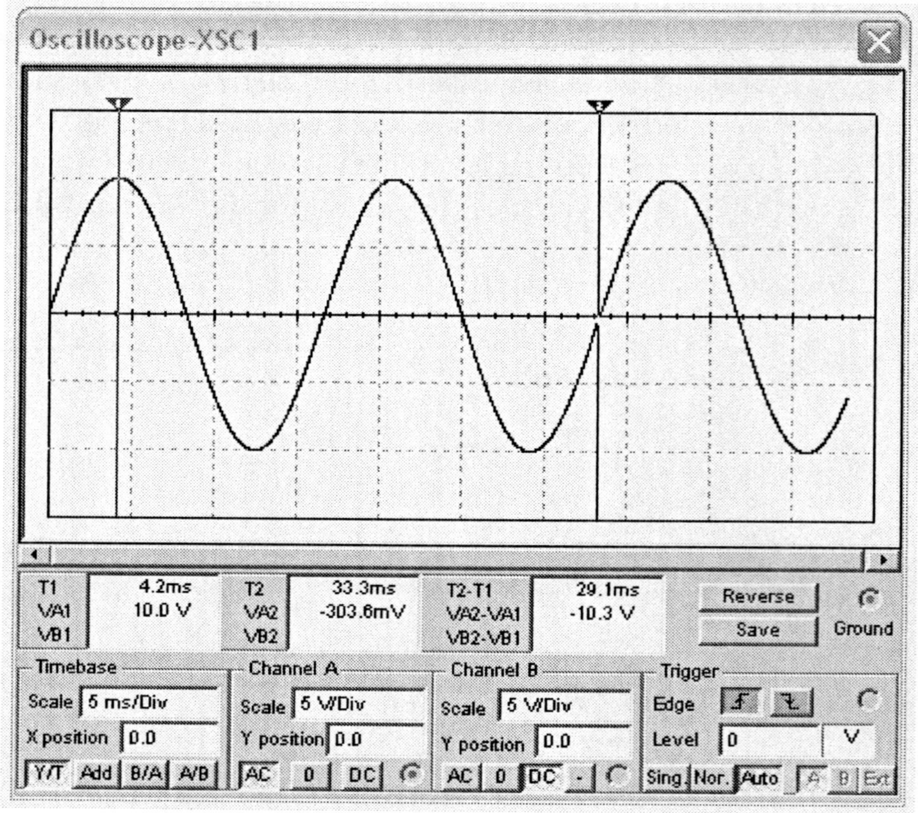

Answer

NAME ____________________________________ DATE __________________

21-3. Open the Multisim circuit file CH21 _03 shown. Turn the simulation on and double click the oscilloscope and multimeter icons to view the display and front panel controls.

(a) Which instrument measures rms values?
(b) What is the rms value of the AC voltage?
(c) Which instrument measures peak and peak-to-peak values?
(d) What is the peak value of the AC voltage?
(e) What is the peak-to-peak value of the AC voltage?

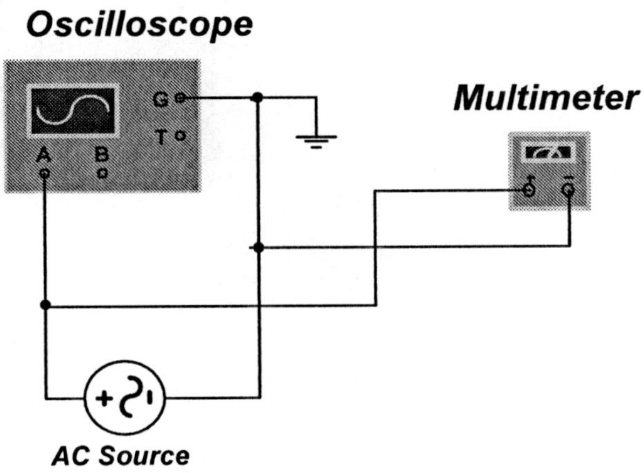

Answers

NAME ______________________________ DATE ________________

21-4 Open the Multisim circuit file CH21 _04 shown and turn the simulation on.

(a) Double click the AC input icon and record the setting of the peak amplitude of the AC sine wave.

(b) Double click the scope input icon and draw the shape of the DC voltage seen.

(c) Calculate the average value of the DC output voltage.

(d) Connect the multimeter and record the value of the DC output voltage.

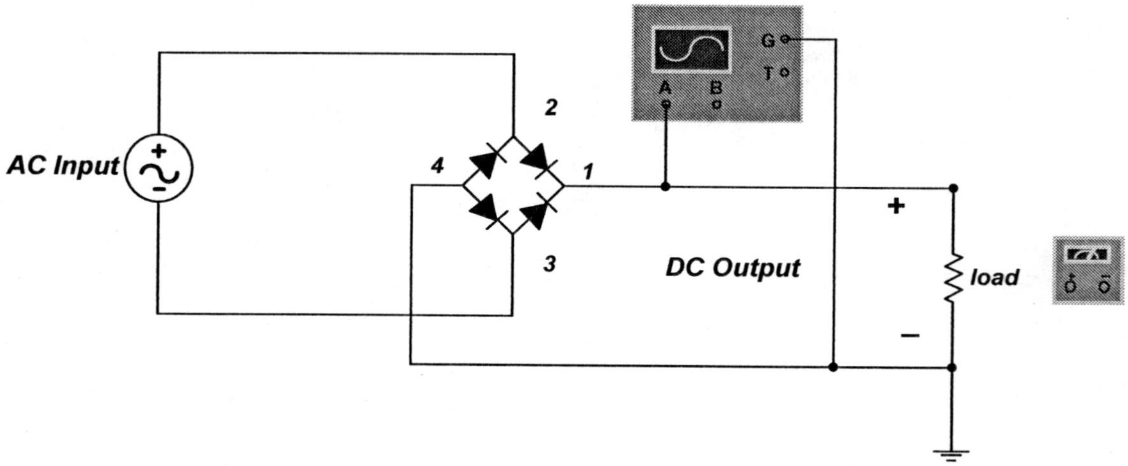

Answers

NAME ______________________________________ DATE __________________

21-5 Open the Multisim circuit file CH21 _05 shown. Connect the meter and record the value of the:
(a) High-voltage primary
(b) Black (Hot) to White (Neutral)
(c) Red (Hot) to White (Neutral)
(d) Black (Hot) to Red (Hot)

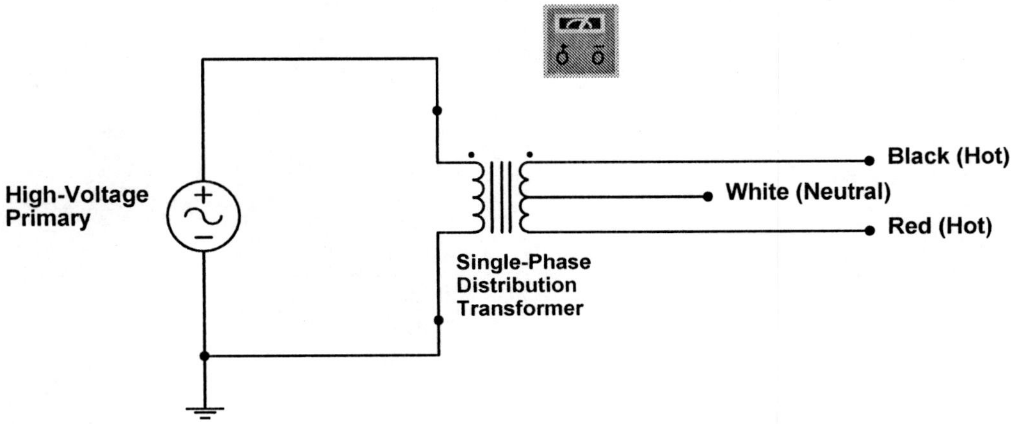

Answers

NAME ______________________________ DATE ________________

21-6 Open the Multisim circuit file CH21 _06 shown and turn the simulation on.

(a) Double click the scope icon and exam Phase "A" and Phase "B" waveforms. How many electrical degrees separate the start of Phase "B" from Phase "A"?

(b) Use the multimeter to record the value of the voltage from Phase "A" to Neutral.

(c) Calculate the line voltage from "A" to "B".

(d) Measure the value of the line voltage from "A" to "B".

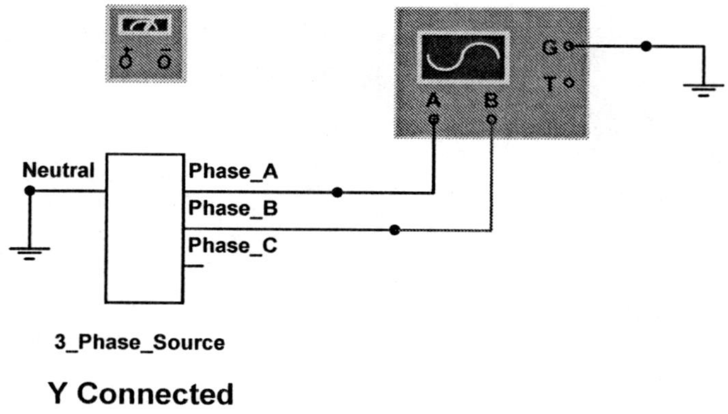

Answers

NAME ______________________________________ DATE __________________

21-7 Open the Multisim circuit file CH21 _07 shown and turn the simulation on.

(a) Use the multimeter to record the value of the line voltage from A-to-B, A-to-C, and B-to-C.

(b) Calculate the phase voltage from line "A" , "B" , or "C" to Neutral.

(c) Measure the value of the phase voltage from line "A" , "B" , or "C" to Neutral.

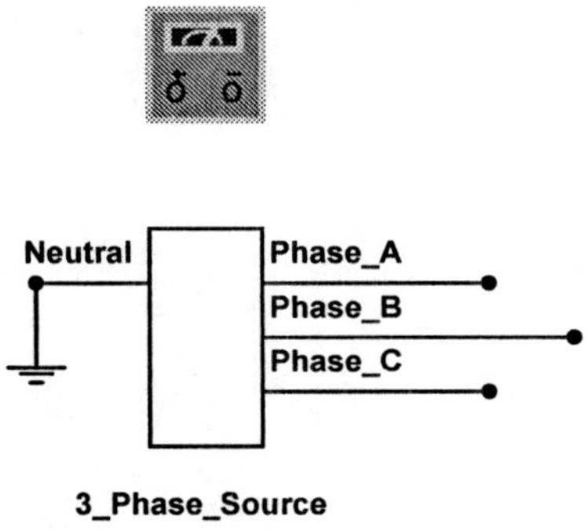

Y Connected

Answers

NAME ________________________________ DATE ________________

21-8 Open the Multisim circuit file CH21 _08 shown and turn the simulation on.

(a) Double click the scope input icon and sketch the shape of the DC voltage waveform seen.

(b) Use the multimeter to record the value of the ***AC*** alternator output voltage from Phase "A" to Ground.

(c) Use the multimeter to record the value of the average ***DC*** output voltage from the POS of the battery to GND.

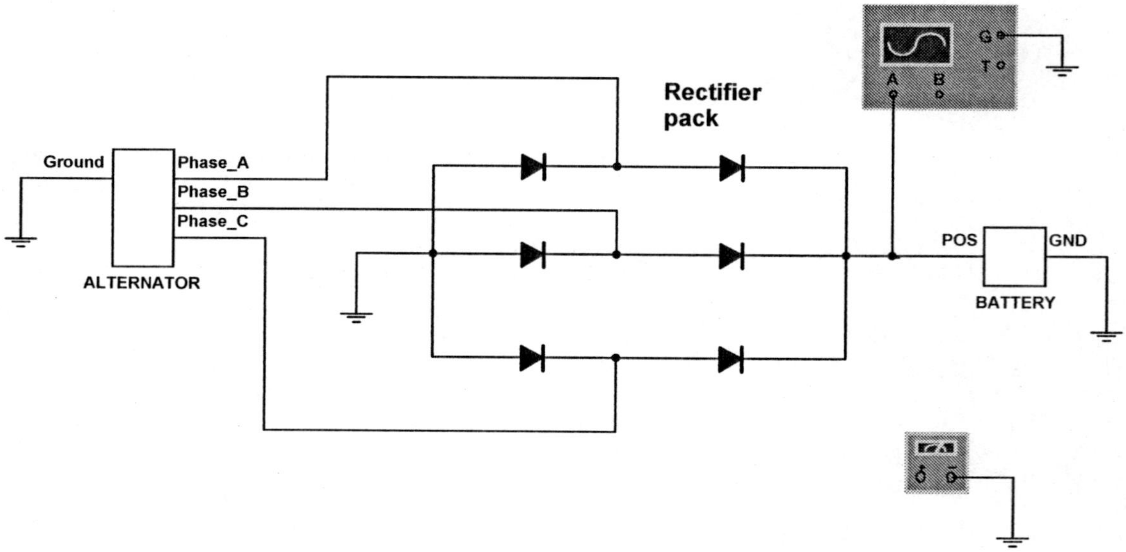

Answers

NAME ______________________________ DATE ________________

21-9. (a) Open the Multisim circuit file CH21 _09 shown. Find all the unknown values of E, I, and R, for the AC resistive series circuit. Record the all given and calculated values in the table provided.
(b) Use the multimeter to measure all values of E, I, R, and record your readings in the table provided.

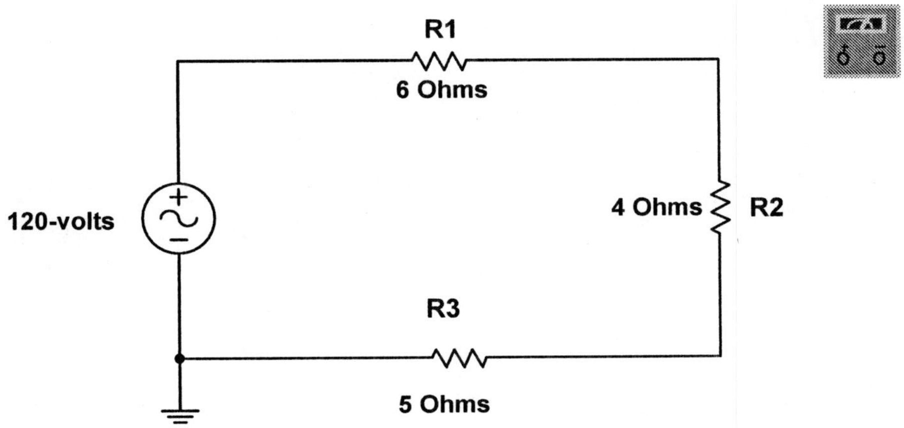

Answers

(a)

	Voltage	Current	Resistance
R_1			
R_2			
R_3			
Total			

(b)

	Voltage	Current	Resistance
R_1			
R_2			
R_3			
Total			

NAME ____________________ DATE ____________

21-10. (a) Open the Multisim circuit file CH21 _10 shown. Find the effective total power consumed by the circuit.
(b) Turn the simulation on and record the total power consumed as indicated by the wattmeter reading.

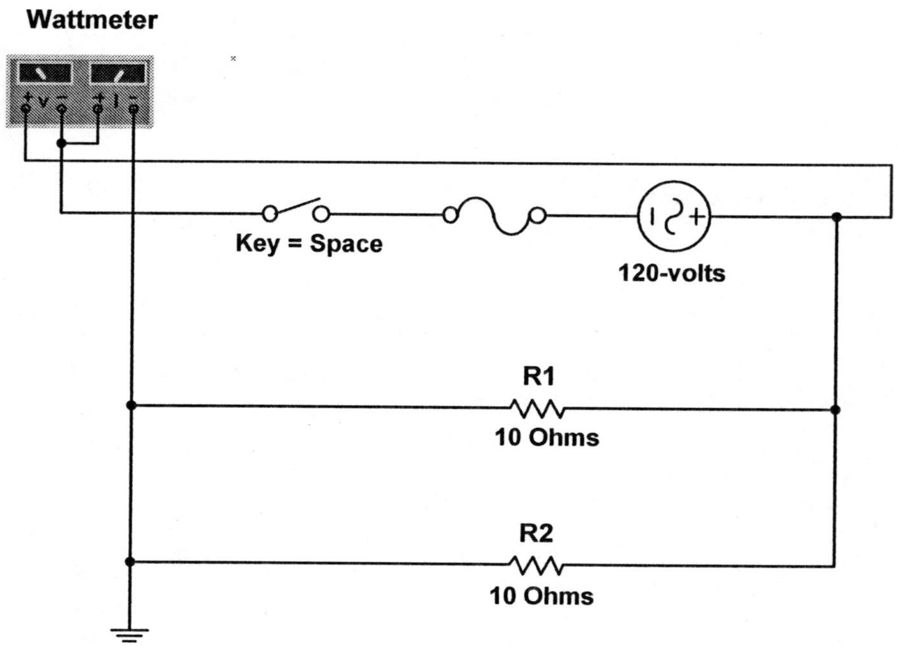

Answers

NAME ____________________ DATE ____________

21-11 Open the Multisim circuit file CH21 _11 shown and turn the simulation on.

(a) Use the multimeter to record the value of the line current and voltage.

(b) Use the multimeter to record the value of the phase current and voltage.

(c) Calculate the total power based on the line current and voltage readings.

(d) Calculate the total power based on the phase current and voltage readings.

(e) Double click the wattmeter icon and record the reading.

(f) What power is the wattmeter reading?

(g) How can the total power be determined from the wattmeter reading?

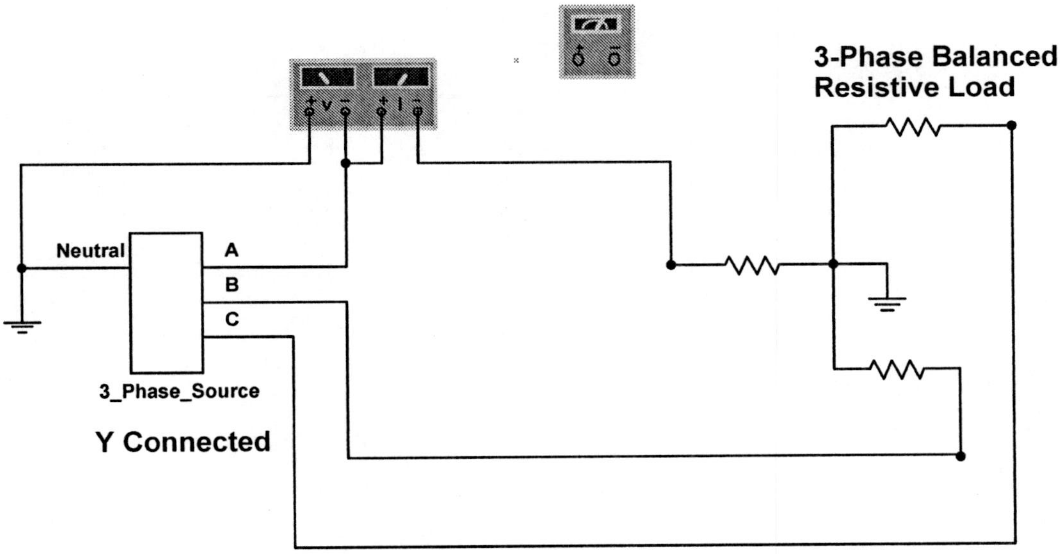

Answers

NAME ______________________ DATE ______________

21-12 Open the Multisim circuit file CH21 _12 shown and turn the simulation on.

(a) Record the value of the current and voltage.

(b) Calculate the total power based on the current and voltage readings.

(c) Double click the wattmeter icons and record the readings W_1 and W_2.

(d) How is the total power be determined from the two wattmeter reading?

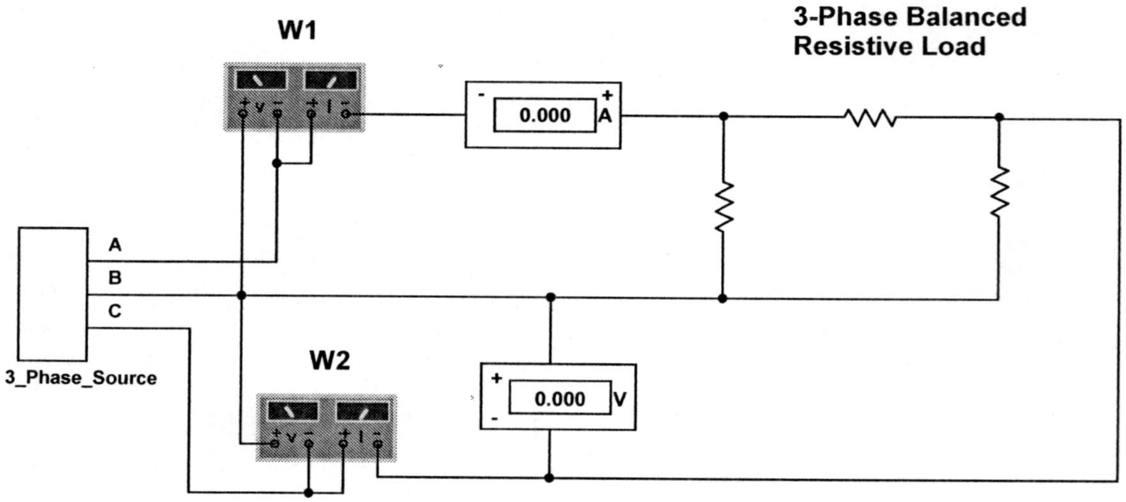

Answers

NAME ______________________________ DATE ____________________

UNIT 22

INDUCTANCE AND CAPACITANCE

Multisim Simulation Assignments

22-1. Open the Multisim circuit file CH22 _01 shown.

(a) Calculate the value of the time constant.

(b) Turn the simulation on, and operate the switch to open and close the circuit to the coil. Sketch the voltage waveform, as displayed by the scope, for the circuit on first closing and opening.

(c) Calculate the time constant when the value of the inductance is 10 H and the value of the resistance remains at 100 Ω.

(d) Double click on the inductor and change its inductance value to 10 H. Operate the simulation and make note of the relative time period it takes for the to reach its maximum and zero values. Explain the reason for this time change.

(e) Change the inductance value to zero to create a purely resistive circuit. Operate the simulation and make note of the relative time period it takes for the to reach its maximum and zero values. Explain the reason for this time change.

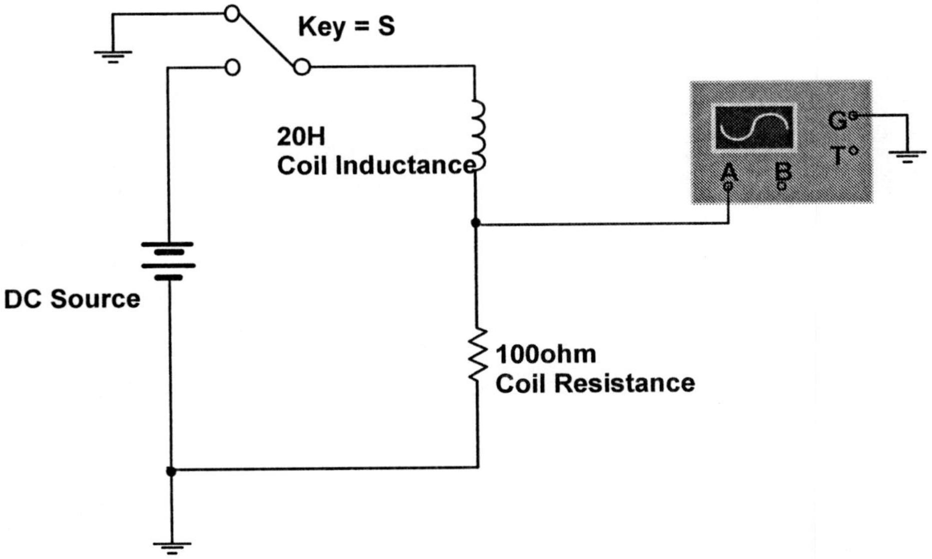

Answers

NAME ______________________________ DATE ________________

22-2. Open the Multisim circuit file CH22 _02 shown and turn the simulation on.

(a) Record the value of the current with the 12 VDC source powering the circuit.

(b) Record the value of the current with the 12 VAC source powering the circuit.

(c) What type of load (resistive - inductive - capacitive) is the lamp classified as?

(d) What type of load (resistive - inductive - capacitive) is the coil classified as?

(e) Which power source produces the greatest amount of current? Why?

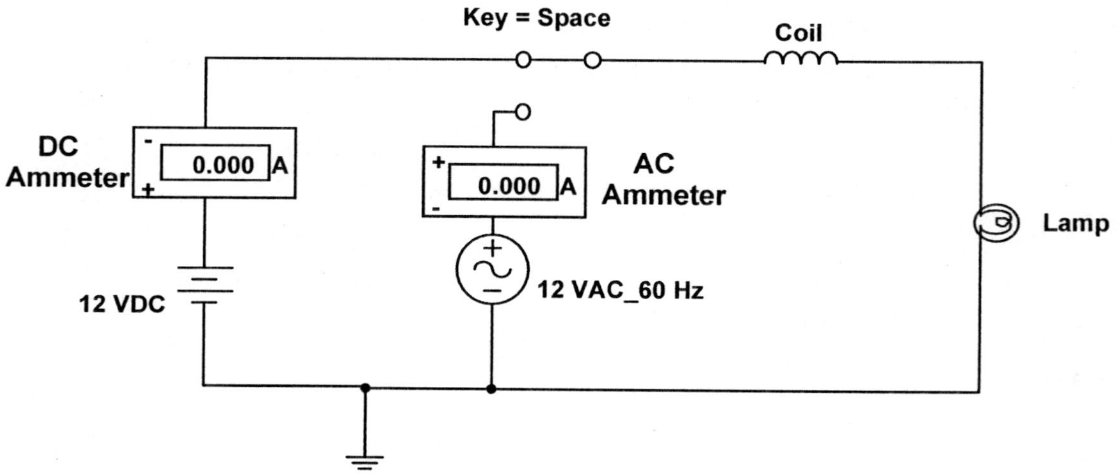

Answers

NAME ________________________________ DATE ________________

22-3. Open the Multisim circuit file CH22 _03 shown and turn the simulation on.

(a) Record the value of the current with the 5 mH coil in series with the lamp.

(b) Record the value of the current with the 10 mH coil in series with the lamp.

(c) Which inductor connection produces the greatest amount of current? Why?

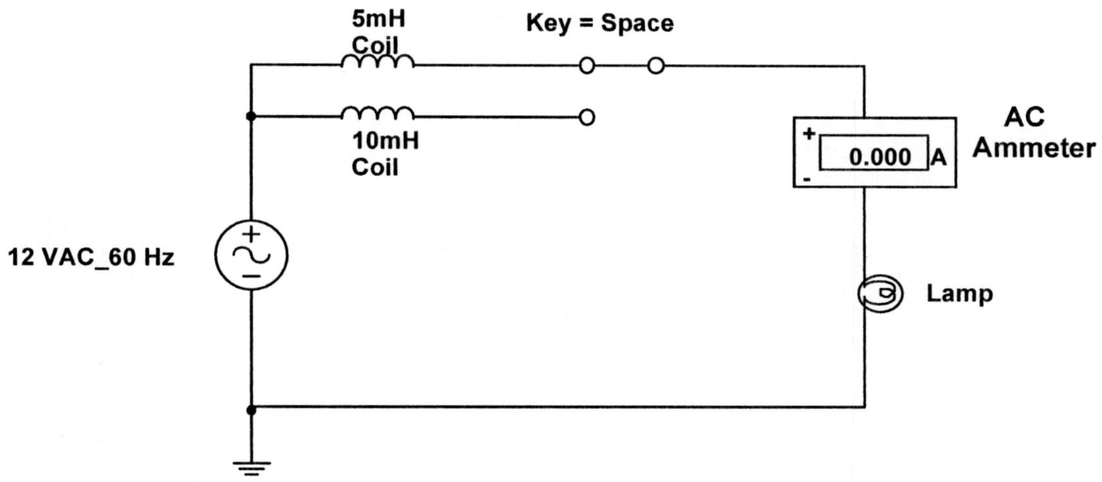

Answers

NAME ________________________________ DATE ____________________

22-4. Open the Multisim circuit file CH22 _04 shown and turn the simulation on.

(a) Record the value of the current with the 60 Hz source powering the circuit.

(b) Record the value of the current with the 300 Hz source powering the circuit.

(c) Which power source produces the greatest amount of current? Why?

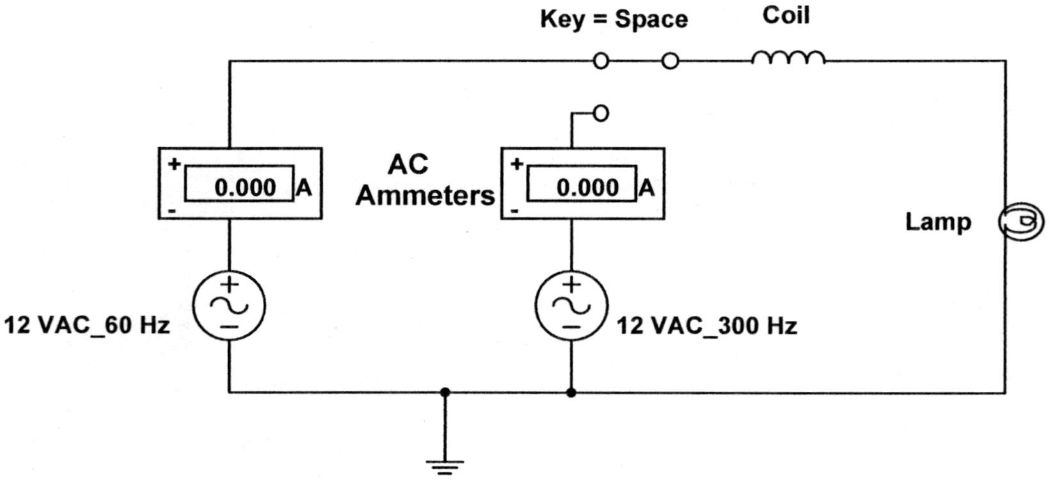

Answers

NAME ______________________________________ DATE ___________________

22-5. Open the Multisim circuit file CH22 _05 shown and turn the simulation on.

(a) Use the multimeter to measure and record the DC resistance of the ideal or pure inductor "L".

(b) Calculate the inductive reactance of the inductor connected to the circuit shown.

(c) Calculate the value of the circuit current flow.

(d) Use the multimeter to measure and record the value of the current flow.

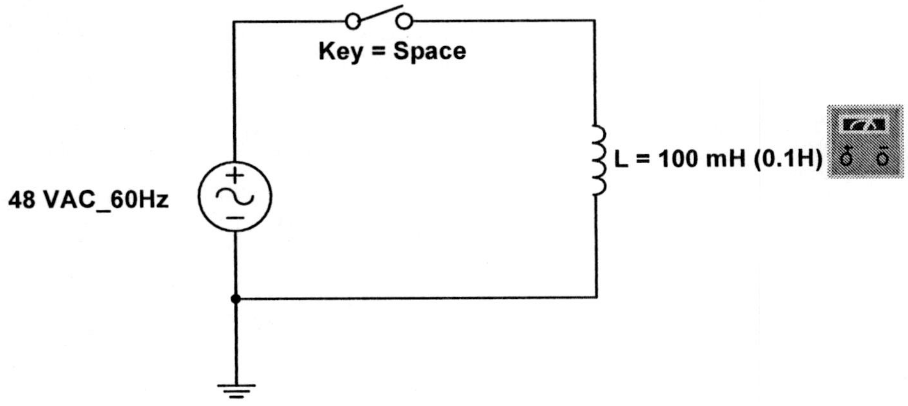

Answers

NAME ________________________________ DATE ____________________

22-6. Open the Multisim circuit file CH22 _06 shown.

(a) Calculate the value of the total circuit inductance.

(b) Calculate the value of the total circuit inductive reactance.

(c) Calculate the value of the circuit current flow.

(d) Turn the simulation on and use the multimeter to measure and record the value of the current flow.

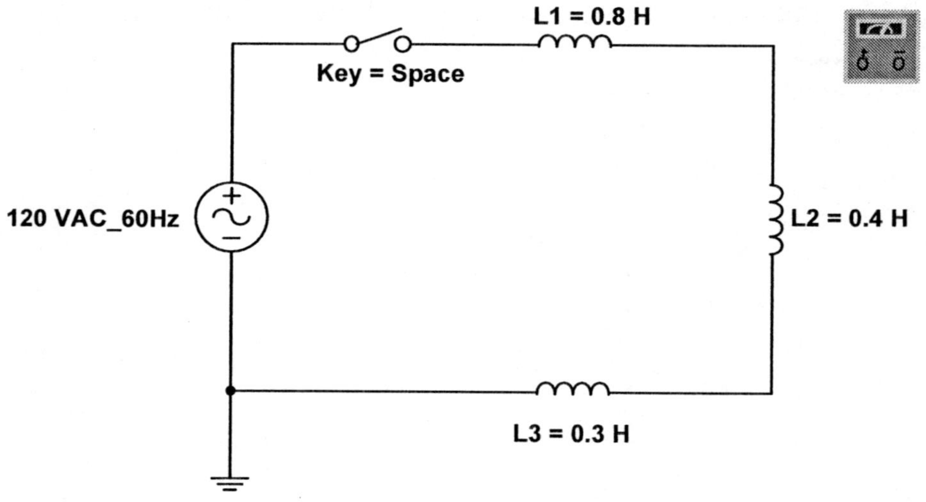

Answers

NAME ______________________________ DATE ________________

22-7. Open the Multisim circuit file CH22 _07 shown.

(a) Calculate the value of the total circuit inductance.

(b) Calculate the value of the total circuit inductive reactance.

(c) Calculate the value of the total circuit current flow.

(d) Turn the simulation on and use the multimeter to measure and record the value of the total current.

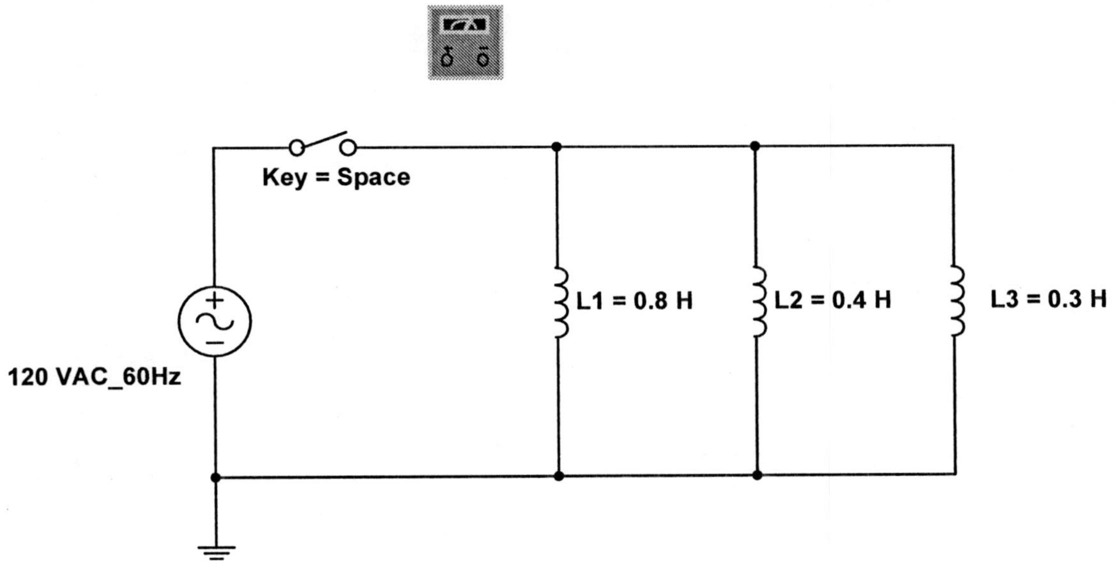

Answers

NAME ______________________________ DATE ________________

22-8. Open the Multisim circuit file CH22 _08 shown. This simulation is designed to show the current and voltage phase relationship for pure resistive and pure inductive loads. The red trace of the scope represents the voltage and the blue trace represents the current. Turn the simulation on and record the following observations:

(a) Phase relationship between current and voltage with the switch set to the "resistor position".

(b) Phase relationship between current and voltage with the switch set to the "inductor position".

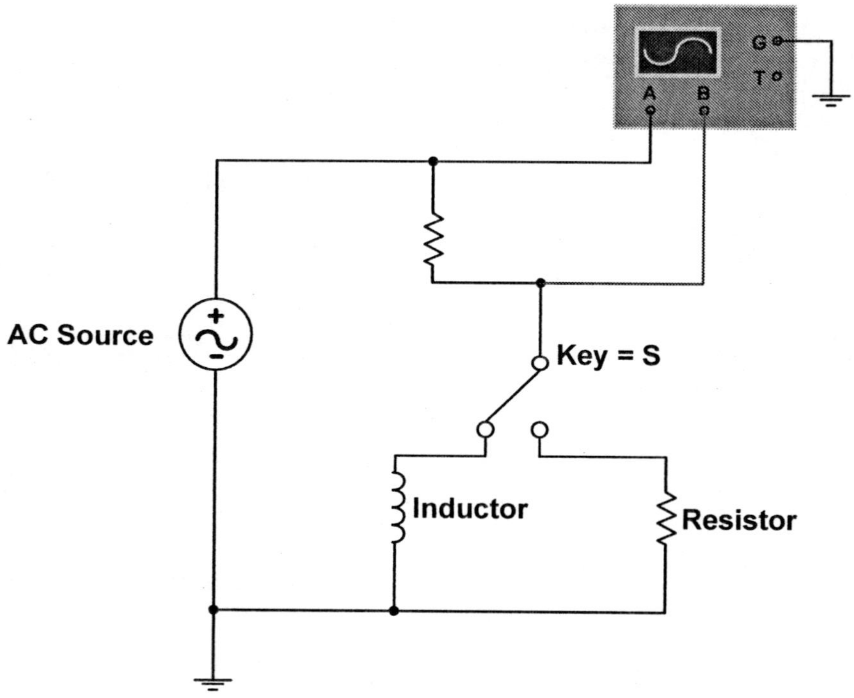

Answers

NAME ______________________________ DATE ________________

22-9. Open the Multisim circuit file CH22 _09 shown and turn the simulation on.

(a) Record the value of the voltage, current, and wattage of the circuit with the resistive load connected.

(b) Record the value of the voltage, current, and wattage of the circuit with the inductive load connected.

(c) Which measurement is noticeably different for the two components? Why?

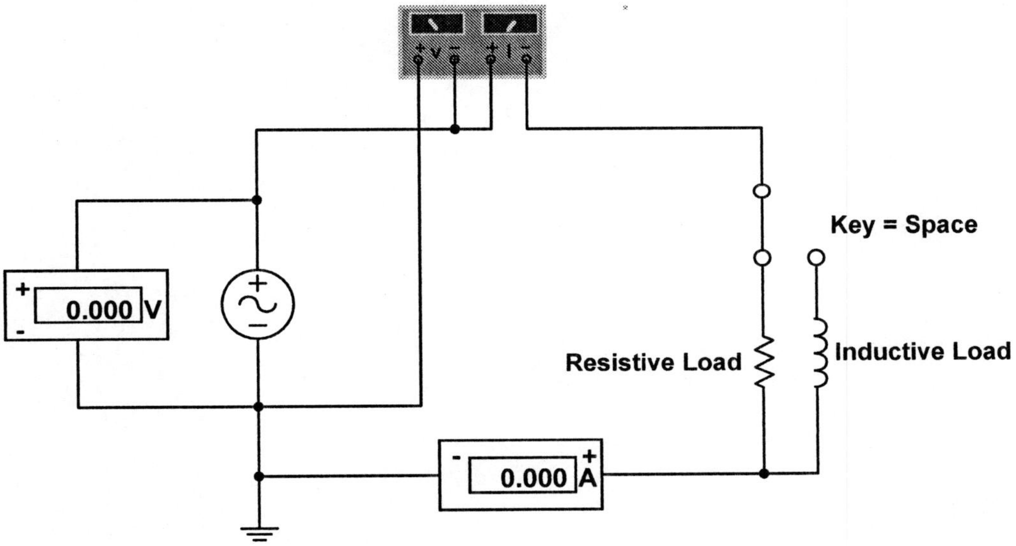

Answers

NAME ________________________________ DATE ____________________

22-10. Open the Multisim circuit file CH22 _10 shown.

(a) Calculate the value of the inductive reactance.

(b) Calculate the value of the current flow.

(c) Turn the simulation on and use the multimeter to measure and record the value of the current flow.

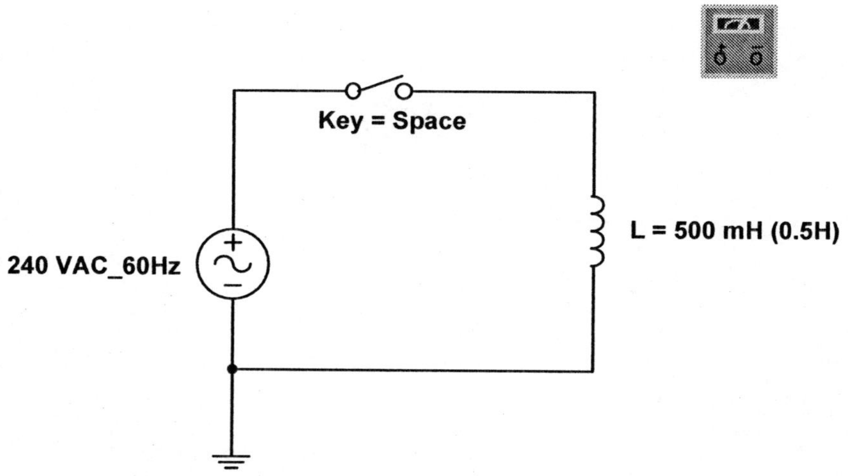

Answers

NAME ________________________________ DATE ________________

22-11. Open the Multisim circuit file CH22 _11 shown.

(a) Calculate the value of the total inductance.

(b) Calculate the value of the total inductive reactance.

(c) Calculate the value of the current flow.

(d) Turn the simulation on and use the multimeter to measure and record the value of the current flow.

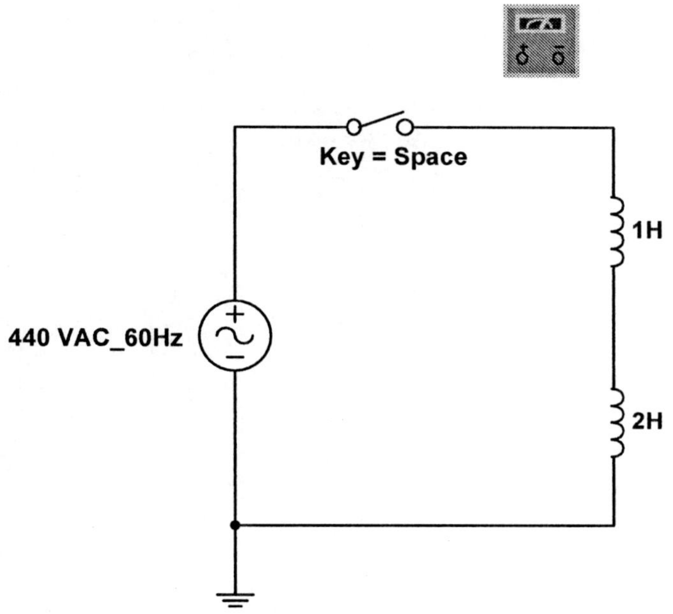

Answers

NAME ______________________________________ DATE ____________________

22-12. Open the Multisim circuit file CH22 _12 shown.

(a) Calculate the value of the total inductance.

(b) Calculate the value of the total inductive reactance.

(c) Calculate the value of the current flow.

(d) Turn the simulation on and use the multimeter to measure and record the value of the current flow.

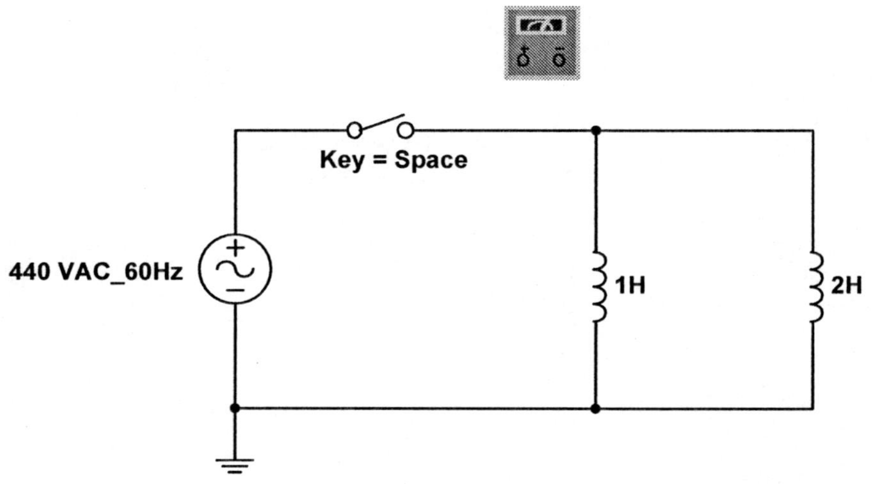

Answers

NAME ______________________________ DATE ________________

22-13. Open the Multisim circuit file CH22 _13 shown.

(a) Calculate the value of the inductive reactance.

(b) Determine the value of the ammeter reading.

(c) Determine the value of the wattmeter reading.

(d) Turn the simulation on and record the value of the ammeter and wattmeter readings.

(e) Calculate the value of the VARs.

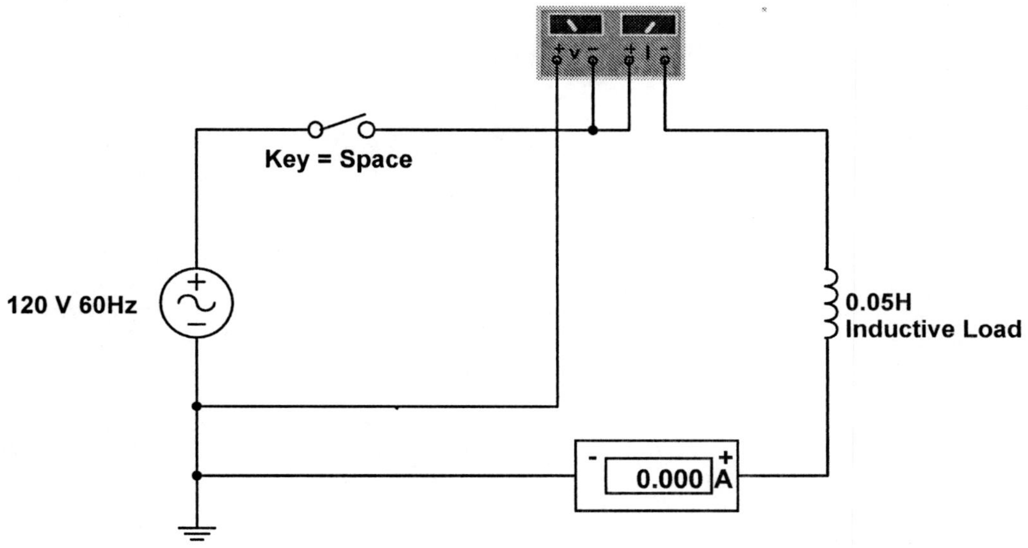

Answers

NAME ______________________________ DATE ______________

22-14. Open the Multisim circuit file CH22 _14 shown. This simulation is designed to show the charging and discharging action of a capacitor.

(a) Turn the simulation on and operate the switch to fully charge the capacitor. What indication do you have that the capacitor is fully charged?

(b) Once charged, operate the switch to discharge the capacitor through the lamp. Compare the direction of the charging and discharging current flow as indicated by the ammeter.

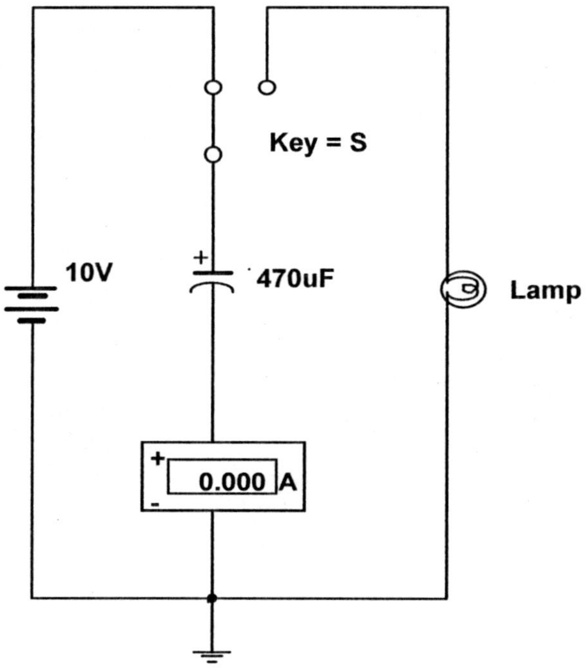

Answers

NAME ______________________________ DATE ______________

22-15. Open the Multisim circuit file CH22 _15 shown and turn the simulation on.

(a) Apply DC directly to the capacitor and make note of the relative time it takes for the capacitor to charge to the value of the source voltage.

(b) Repeat with the 1 kΩ resistor connected in series with the resistor.

(c) Repeat with the 10 kΩ resistor connected in series with the resistor.

(d) How does the resistance value of the affect the time it takes for the capacitor to charge to the maximum value?

(e) The higher the resistance the longer the charging time.

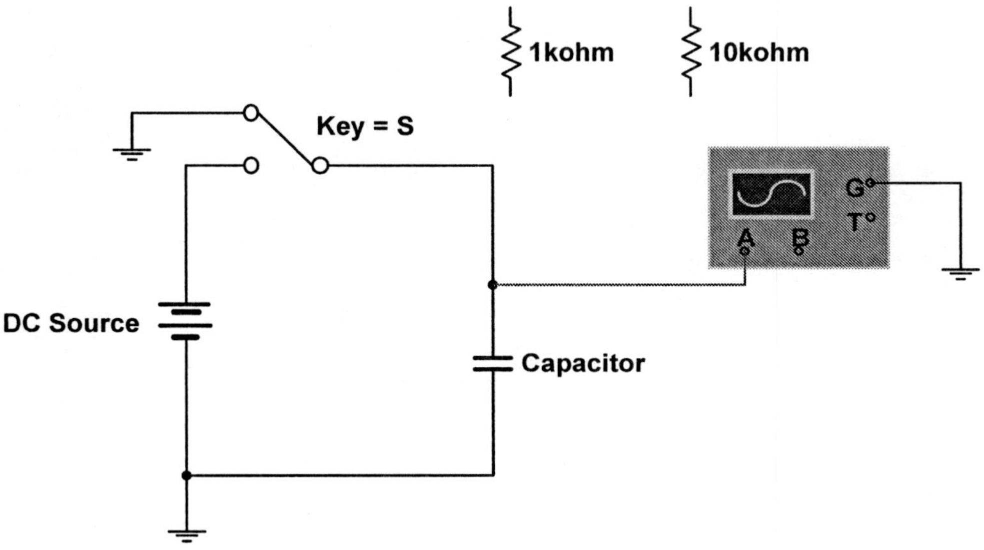

Answers

NAME ______________________________ DATE ______________

22-16. Open the Multisim circuit file CH22 _16 shown.

(a) Calculate the value of the RC time constant.

(b) Approximately how long will it take the capacitor to fully charge to 12 volts?.

(c) Approximately how long will it take the capacitor to discharge to zero volts?.

(d) Turn the simulation on and referring to the scope time base setting of 1s/Div (1 second per division) record the approximate charge and discharge times.

(e) Repeat steps (a) to (d) with the resistor value changed to 50 kΩ.

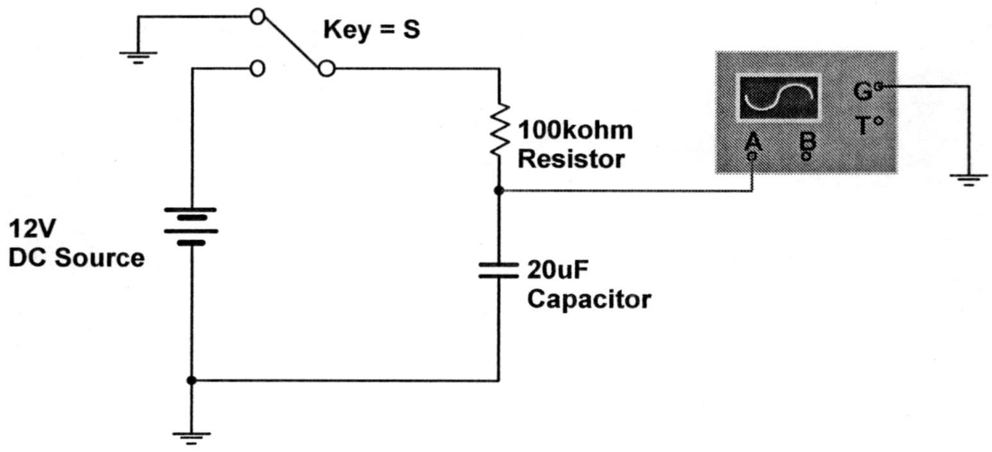

Answers

NAME ____________________________________ DATE ____________________

22-17. Open the Multisim circuit file CH22 _17 shown. Turn the simulation on and operate the circuit. When does the capacitor appear to pass current through the dielectric?

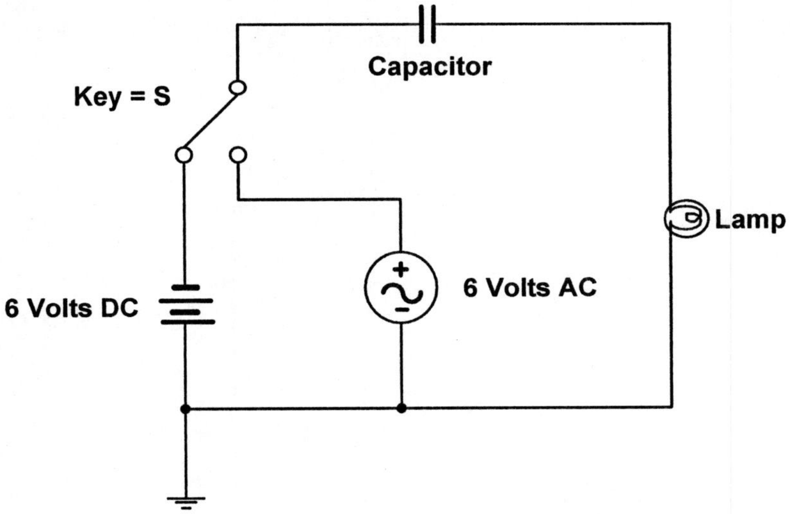

Answer

NAME ____________________ DATE ____________

22-18. Open the Multisim circuit file CH22 _18 shown and turn the simulation on.

(a) Close the switch and record the value of the AC and DC voltage across the lamp.

(b) Record the value of the AC and DC voltage across the capacitor.

(c) Which voltage is blocked by the capacitor from reaching the lamp?

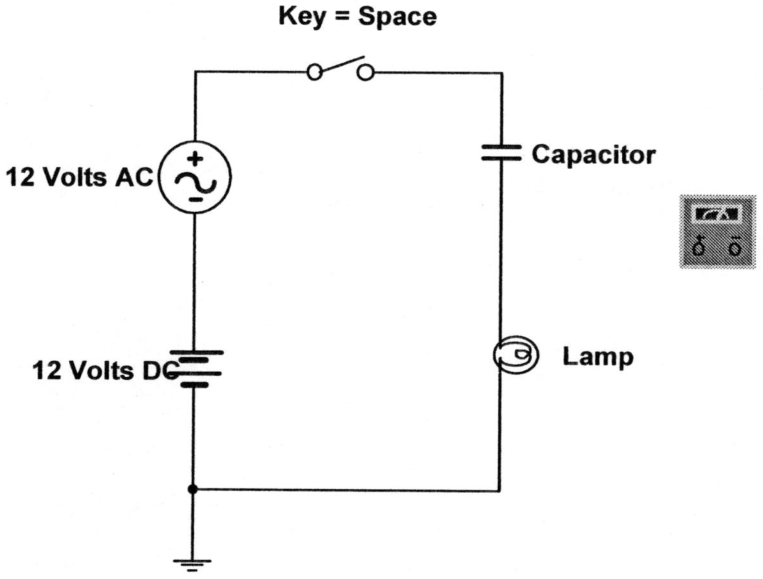

Answers

NAME ______________________________ DATE ________________

22-19. Open the Multisim circuit file CH22 _19 shown and turn the simulation on.

(a) Record the amount of current flow with the 10μF capacitor connected in series with the lamp.

(b) Record the amount of current flow with the 50μF capacitor connected in series with the lamp.

(c) Which capacitor connection resulted in the most current flow? Why?

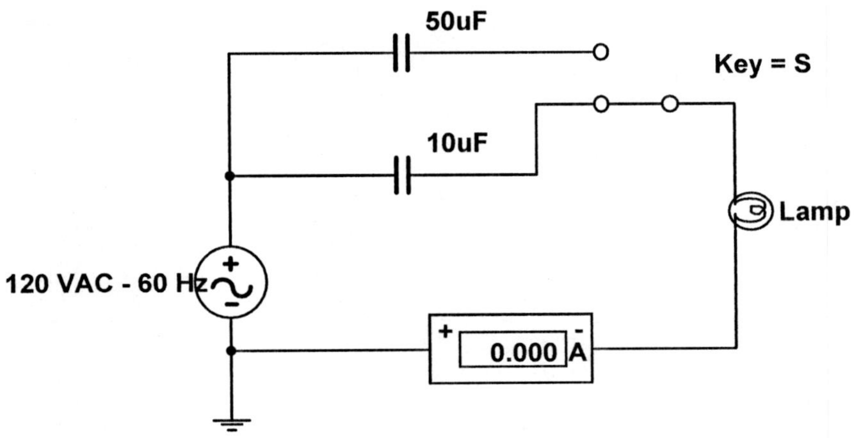

Answers

NAME ______________________________ DATE ________________

22-20. Open the Multisim circuit file CH22 _20 shown and turn the simulation on.

(a) Record the amount of current flow with the 120 V, 60 Hz, power supply connected in series with the lamp.

(b) Record the amount of current flow with the 120 V, 600 Hz, power supply connected in series with the lamp.

(c) Which frequency resulted in the most current flow? Why?

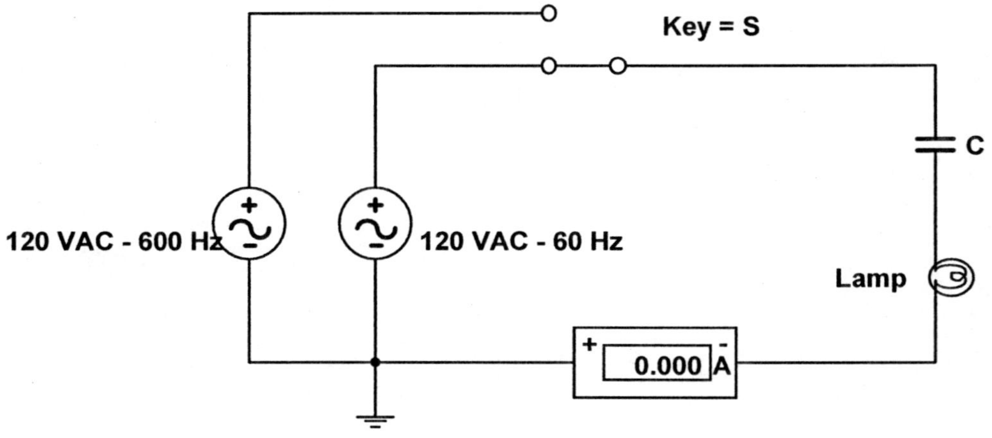

Answers

NAME ______________________________________ DATE ____________________

22-21. Open the Multisim circuit file CH22 _21 shown.

(a) Determine the capacitive reactance of the capacitor.

(b) Calculate the value of the current flow in the circuit.

(c) Measure and record the amount of current flow in the circuit.

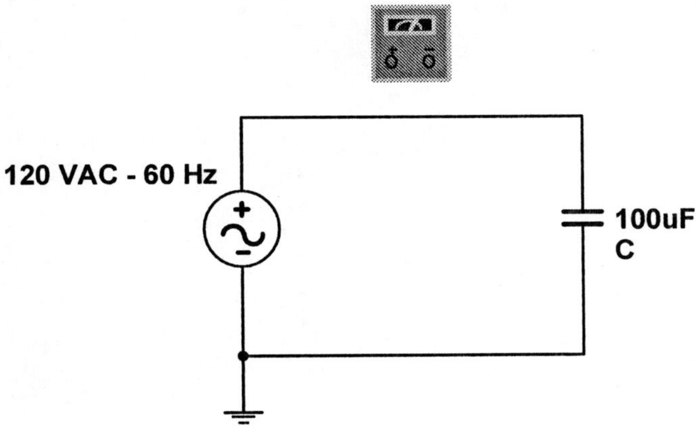

Answers

NAME ______________________________ DATE ________________

22-22. Open the Multisim circuit file CH22 _22 shown.

(a) Calculate the total capacitance of the circuit.
(b) Calculate the total capacitive reactance of the circuit.
(c) Calculate the value of the current flow in the circuit.
(d) Measure and record the amount of current flow in the circuit.
(e) Calculate the value of the voltage drop across each capacitor.
(f) Measure and record the value of the voltage drop across each capacitor.

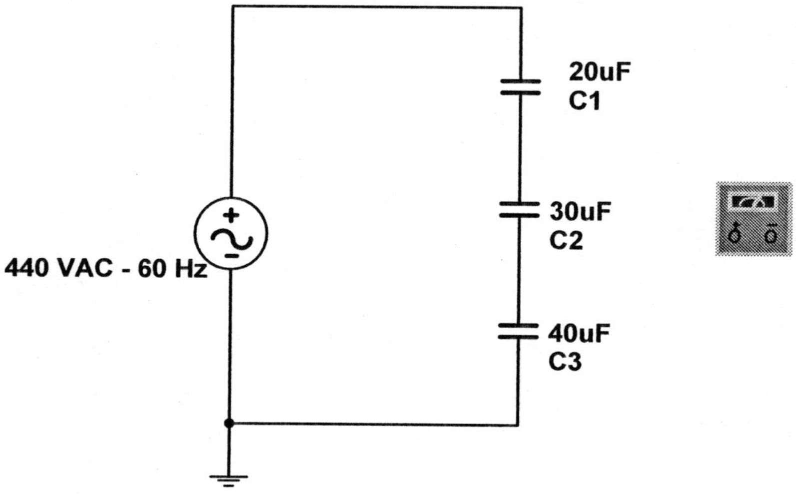

Answers

NAME ______________________________ DATE ________________

22-23. Open the Multisim circuit file CH22 _23 shown.
(a) Calculate the total capacitance of the circuit.
(b) Calculate the total capacitive reactance of the circuit.
(c) Calculate the value of the total current flow of the circuit.
(d) Measure and record the amount of total current flow of the circuit.
(e) Calculate the value of the current flow through each capacitor.
(f) Measure and record the value of the current flow through each capacitor.

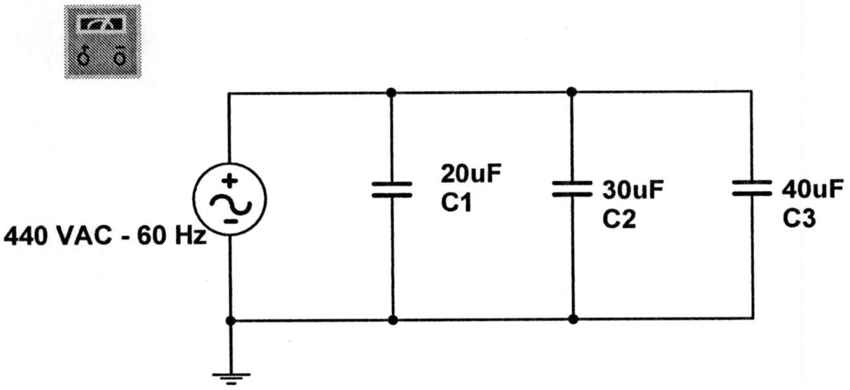

Answers

NAME ____________________ DATE ____________

22-24. Open the Multisim circuit file CH22 _24 shown. This simulation is designed to show the current and voltage phase relationship for a pure capacitive load. The red trace of the scope represents the voltage and the blue trace represents the current. Turn the simulation on and study the current and voltage waveforms relative to each other. What is the phase relationship between the current and the voltage?

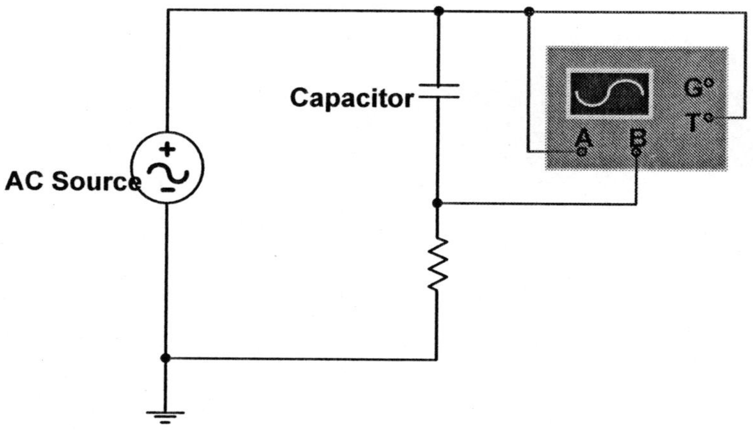

Answer

NAME ______________________________ DATE ________________

22-25. Open the Multisim circuit file CH22 _25 shown and turn the simulation on.

(a) Record the value of the current flow through the inductor and capacitor.

(b) Calculate the value of the total line current based on the fact that these two current are 180° out-of-phase and act to oppose each other.

(c) Measure and record the value of the line current.

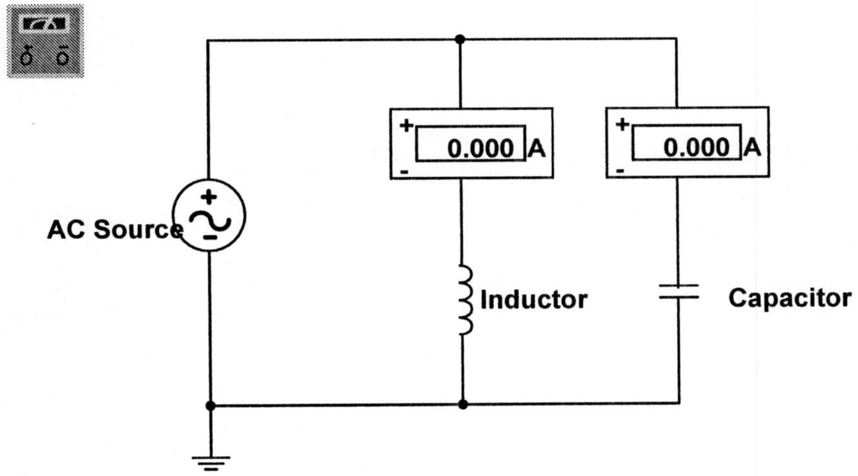

ANSWERS

NAME ______________________________ DATE ________________

22-26. Open the Multisim circuit file CH22 _26 shown and turn the simulation on.

(a) Record the value of the voltage, current, and wattage of the circuit with the resistive load connected.

(b) Record the value of the voltage, current, and wattage of the circuit with the inductive load connected.

(c) Which measurement is noticeably different for the two components? Why?

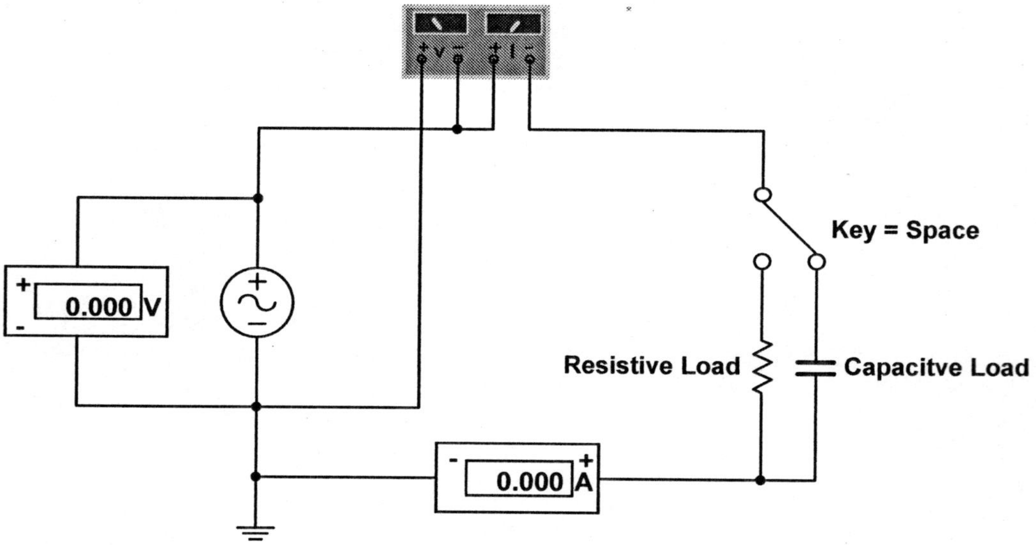

Answers

NAME ______________________ DATE ______________

22-27. Open the Multisim circuit file CH22 _27 shown.

(a) Calculate the capacitance reactance of the capacitor.

(b) Calculate the value of the current flow.

(c) Measure and record the value of the current flow.

Answers

NAME ______________________________ DATE ____________________

22-28. Open the Multisim circuit file CH22 _28 shown.

(a) Calculate the value of the capacitive reactance.

(b) Determine the value of the ammeter reading.

(c) Calculate the value of the VARs.

(d) Determine the value of the wattmeter reading.

(e) Turn the simulation on and record the value of the ammeter and wattmeter readings.

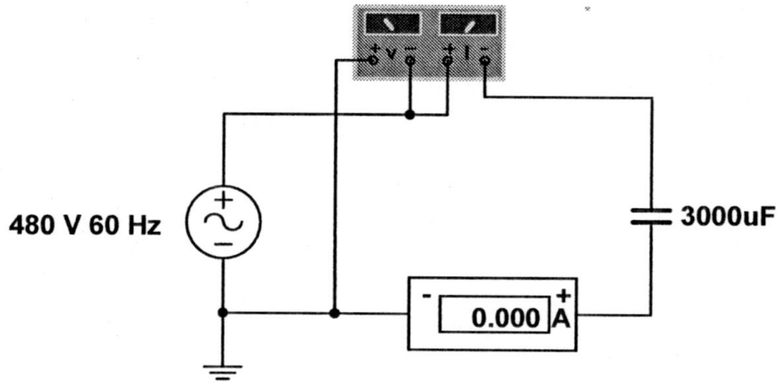

Answers

NAME ________________________________ DATE ________________

UNIT 23

RESISTIVE, INDUCTIVE, CAPACITIVE, (RLC) CIRCUITS IN SERIES

Multisim Simulation Assignments

23-1. Open the Multisim circuit file CH23 _01 shown. Turn the simulation on and double-click the scope icon. Observe voltage waveforms E_A and E_B relative magnitude and phase shift. Construct a vector representation of the two waveforms.

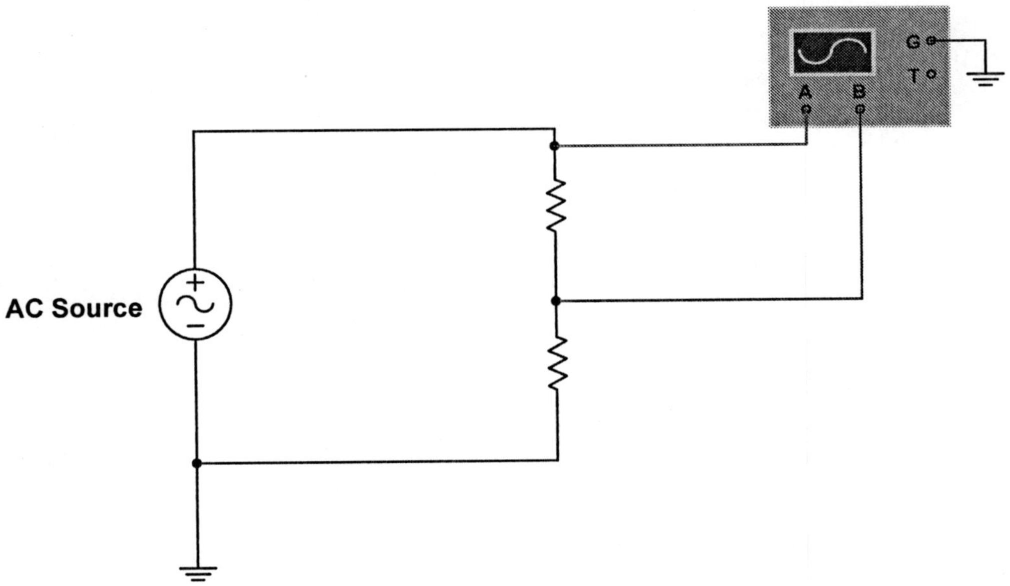

Answer

NAME ______________________________ DATE ________________

23-2. Open the Multisim circuit file CH23 _02 shown. Turn the simulation on and double-click the scope icon. Observe voltage waveforms E_A and E_B relative magnitude and phase shift. Construct a vector representation of the two waveforms.

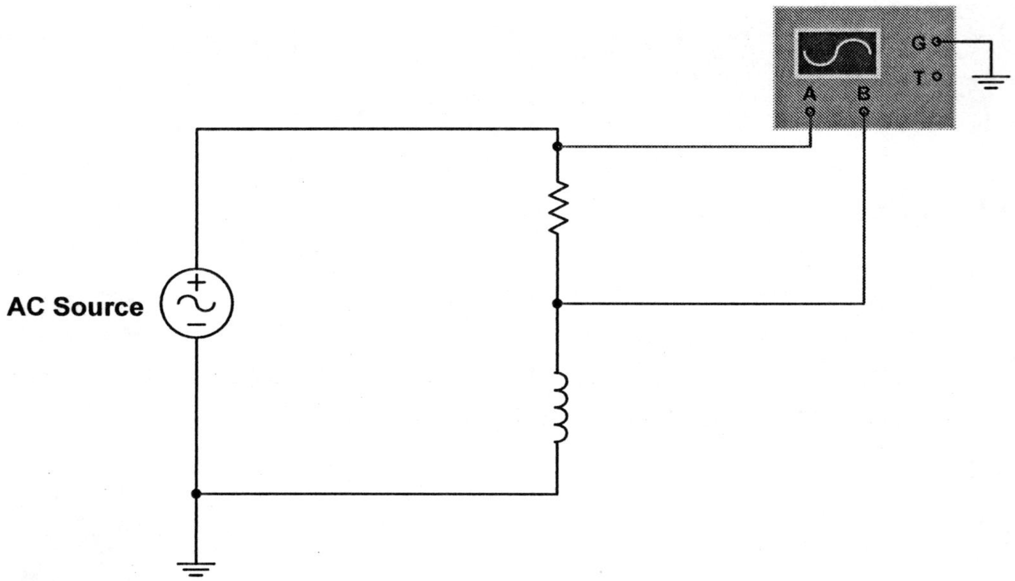

Answer

NAME ________________________________ DATE ________________

23-3. Open the Multisim circuit file CH23 _03 shown.

(a) Calculate the output voltage across the load for circuit # 1 and circuit # 2

(b) Turn the simulation on and measure the output voltage for each circuit.

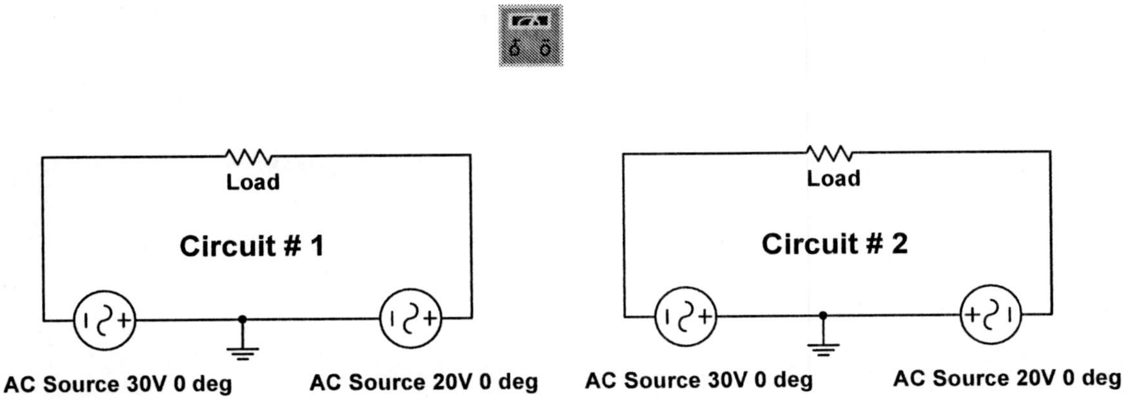

Answers

NAME ______________________ DATE ______________

23-4. Open the Multisim circuit file CH23 _04 shown.

(a) Calculate the output voltage across the load for the circuit shown.

(b) Turn the simulation on and measure the output voltage across the load.

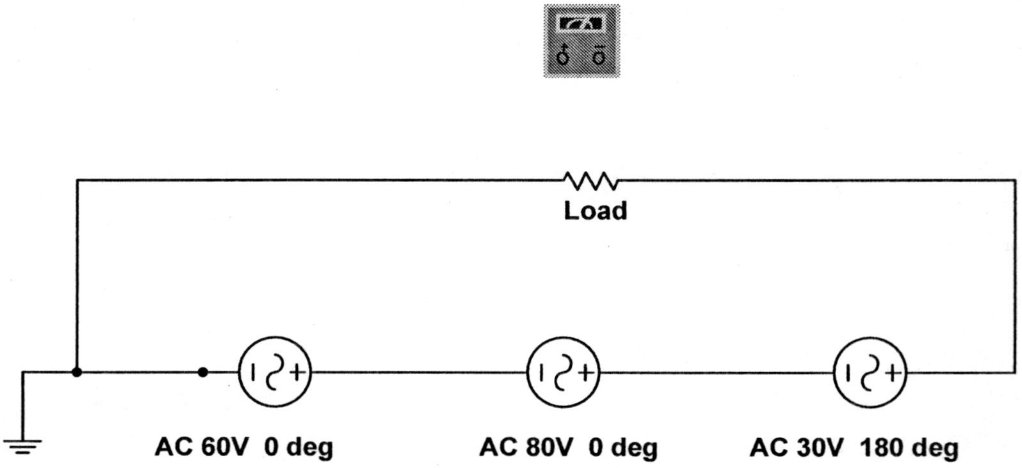

Answers

NAME ______________________________ DATE ________________

23-5. Open the Multisim circuit file CH23 _05 shown.

(a) Determine the output voltage across the load for the circuit shown.

(b) Turn the simulation on and measure the output voltage across the load.

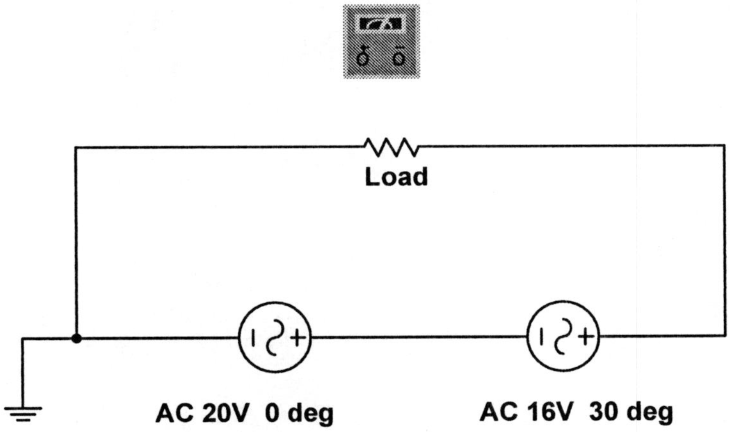

Answers

NAME ______________________________ DATE ________________

23-6. Open the Multisim circuit file CH23 _06 shown.

(a) Determine the output voltage across the load for the circuit shown.

(b) Turn the simulation on and measure the output voltage across the load.

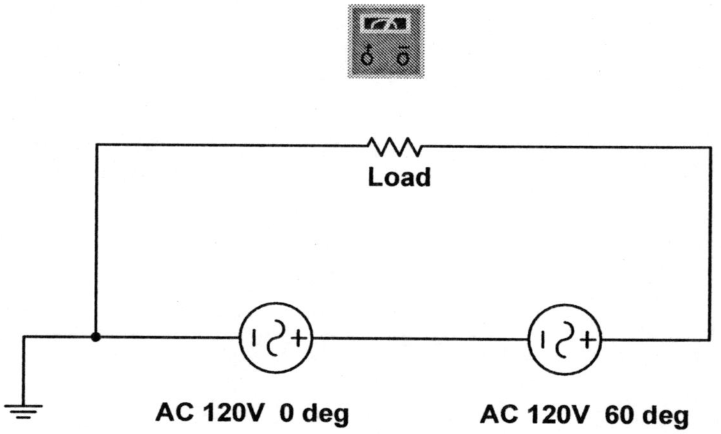

Answers

NAME ____ DATE ____________________

23-7. Open the Multisim circuit file CH23 _07 shown.
(a) Determine the output voltage across the load for the circuit shown.
(b) Turn the simulation on and measure the output voltage across the load.

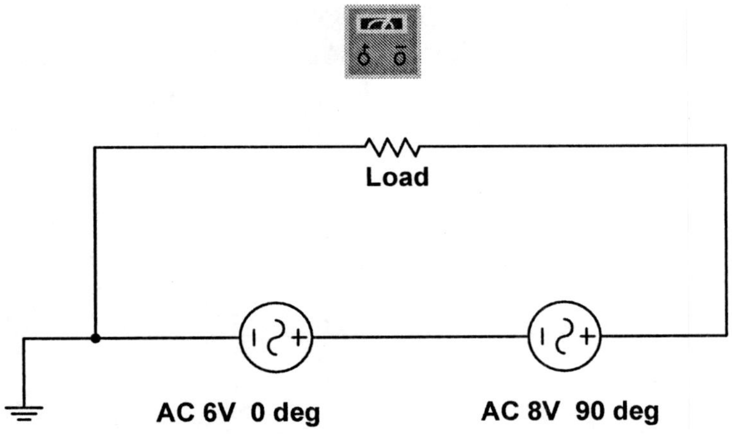

Answers

NAME ______________________ DATE ______________

23-8. Open the Multisim circuit file CH23 _08 shown.

(a) Calculate the impedance for the circuit shown.

(b) Calculate the current flow for the circuit shown.

(c) Connect the multimeter and record the value of the current flow.

(d) Determine the phase angle theta (θ) for the circuit shown.

Inputs "A" and "B" of the scope are connected to show the $\angle\theta$ for the circuit. Double click on the scope icon and examine the phase shift between the two waveforms displayed.

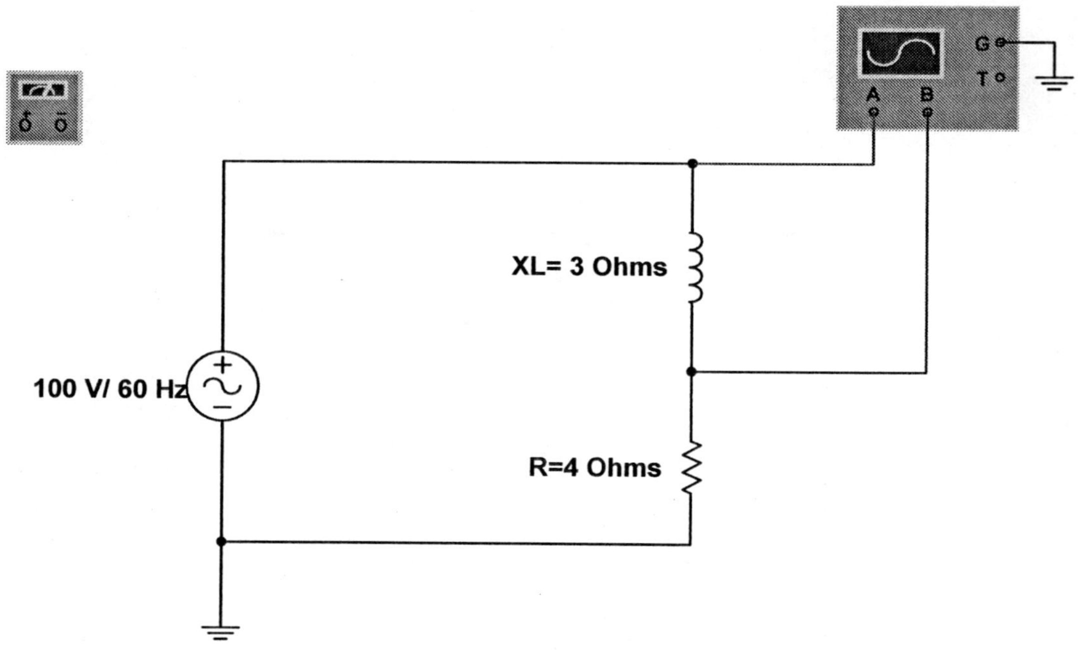

Answers

NAME ____________________ DATE ____________

23-9. Open the Multisim circuit file CH23 _09 shown.

(a) Calculate the current flow for the circuit shown.

(b) Connect the multimeter and record the value of the current flow.

(c) Calculate the value of the voltage drop across the resistor.

(d) Measure the value of the voltage drop across the resistor.

(e) Calculate the value of the voltage drop across the inductor.

(f) Measure the value of the voltage drop across the inductor.

(g) Determine the phase angle theta (θ) for the circuit shown. Inputs "A" and "B" of the scope are connected to show the $\angle\theta$ for the circuit. Double click on the scope icon and examine the phase shift between the two waveforms displayed.

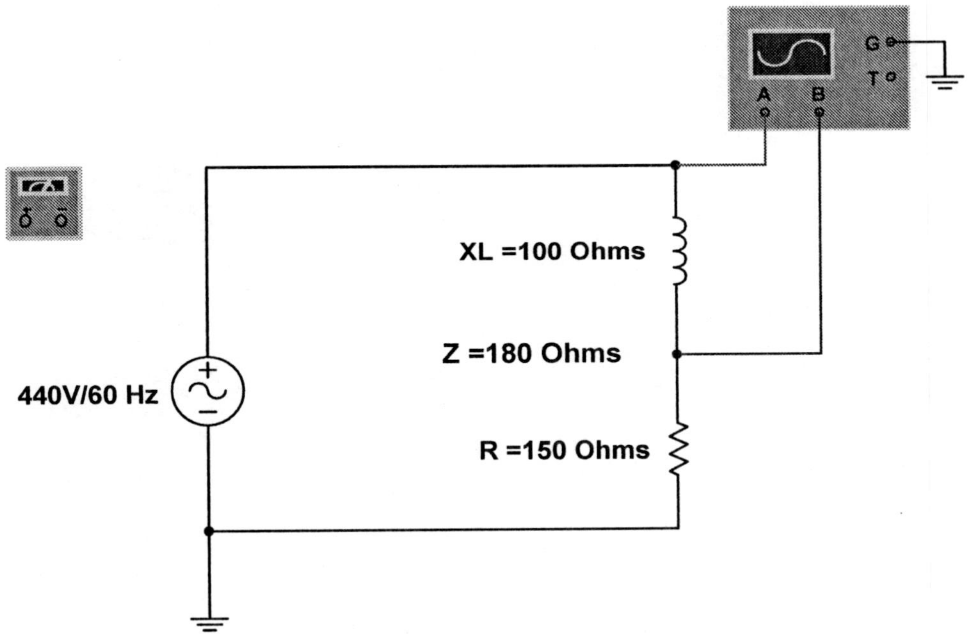

Answers

NAME ________________________________ DATE ________________

23-10. Open the Multisim circuit file CH23 _10 shown.

(a) Record the value of the current (I), applied voltage (E_T), voltage across the resistor (E_R), voltage across the inductor (E_L).

(b) Calculate the value of the apparent power.

(c) Calculate the value of the inductive reactive power.

(d) Calculate the value of the true power.

(e) Calculate the value of the PF

(f) Connect the wattmeter into the circuit to measure the power of the circuit. Record the circuit power and PF as indicated by the wattmeter.

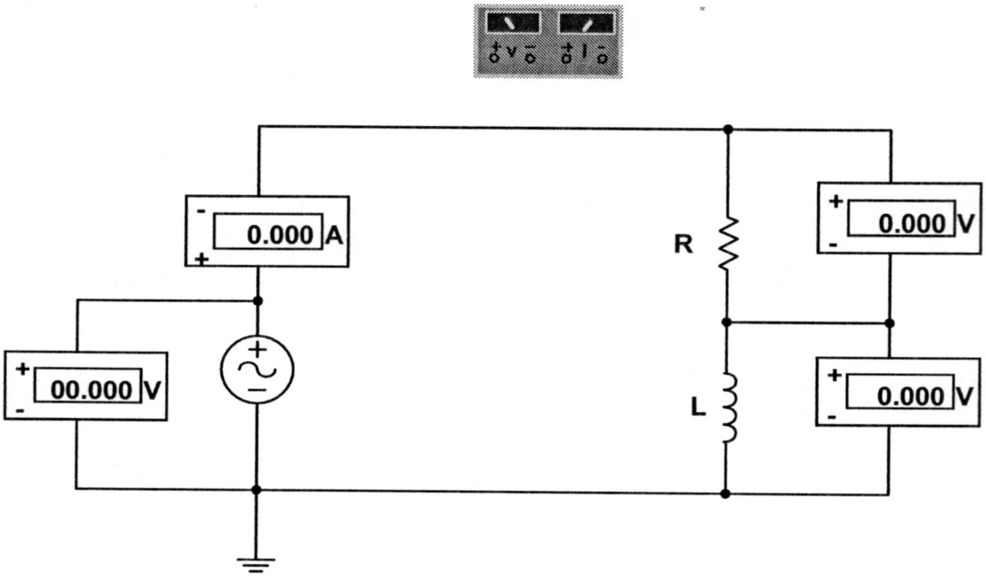

Answers

NAME ______________________________ DATE ________________

23-11. Open the Multisim circuit file CH23 _11 shown.
Use the multimeter, wattmeter, and whatever math calculations may be necessary to complete the table of values for the circuit shown.

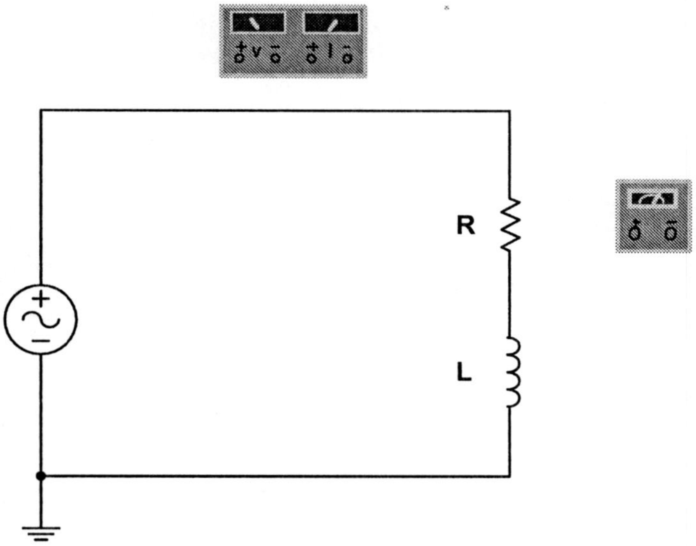

Answers

	E	I	$R\text{-}X_L\text{-}Z$	W-VA-VARs	$\angle\theta$	PF
R						
L						
Total						

NAME ________________________________ DATE ____________________

23-12. Open the Multisim circuit file CH23 _12 shown.

(a) Calculate the impedance for the circuit shown.

(b) Calculate the current flow for the circuit shown.

(c) Connect the multimeter and record the value of the current flow.

(d) Determine the phase angle theta (θ) for the circuit shown.

Inputs "A" and "B" of the scope are connected to show the ∠θ for the circuit. Double click on the scope icon and examine the phase shift between the two waveforms displayed.

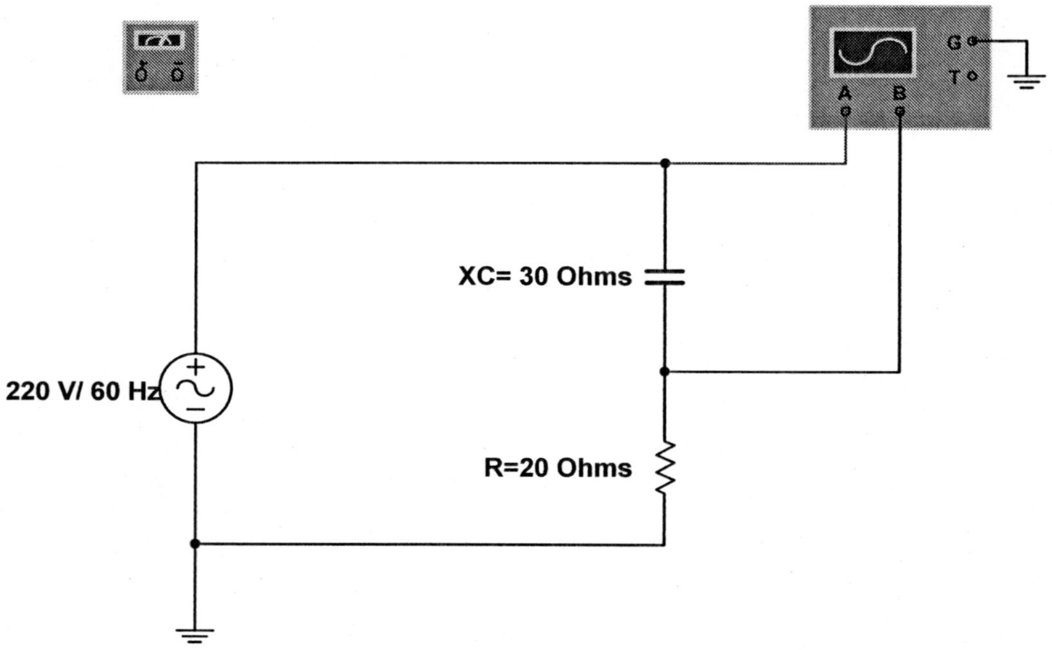

Answers

NAME ______________________ DATE ______________

23-13. Open the Multisim circuit file CH23 _13 shown.

(a) Calculate the current flow for the circuit shown.

(b) Connect the multimeter and record the value of the current flow.

(c) Calculate the value of the voltage drop across the resistor.

(d) Measure the value of the voltage drop across the resistor.

(e) Calculate the value of the voltage drop across the capacitor.

(f) Measure the value of the voltage drop across the capacitor.

(g) Determine the phase angle theta (θ) for the circuit shown. Inputs "A" and "B" of the scope are connected to show the $\angle\theta$ for the circuit. Double click on the scope icon and examine the phase shift between the two waveforms displayed.

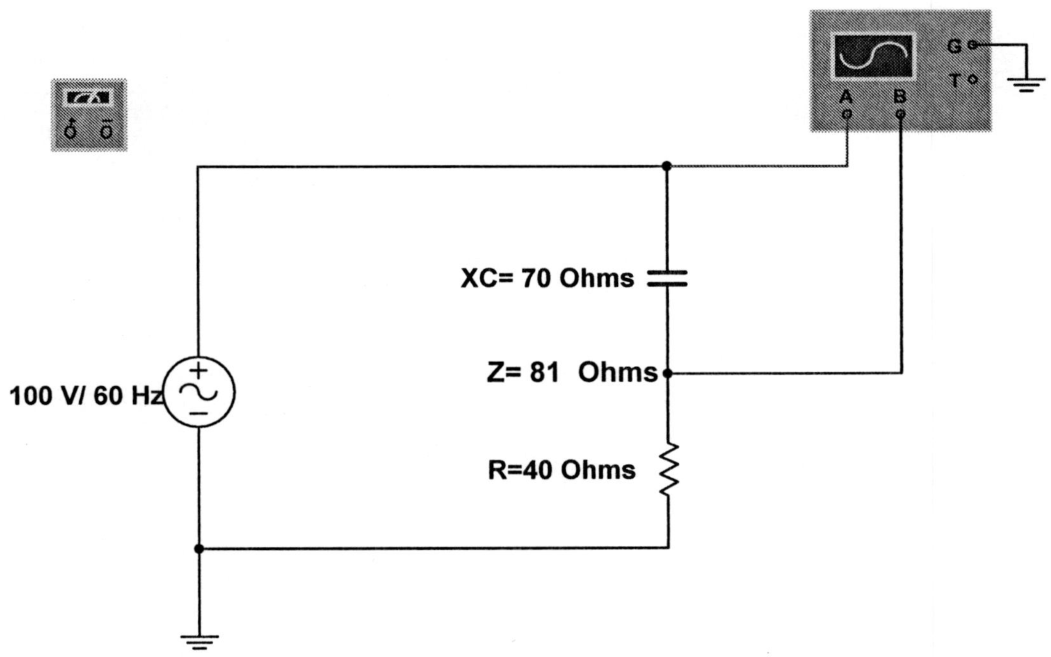

Answers

NAME ____________________ DATE ____________

23-14. Open the Multisim circuit file CH23 _14 shown.

(a) Record the value of the current (I), applied voltage (E_T), voltage across the resistor (E_R), voltage across the inductor (E_L).

(b) Calculate the value of the apparent power.

(c) Calculate the value of the inductive reactive power.

(d) Calculate the value of the true power.

(e) Calculate the value of the PF

(f) Connect the wattmeter into the circuit to measure the power of the circuit. Record the circuit power and PF as indicated by the wattmeter.

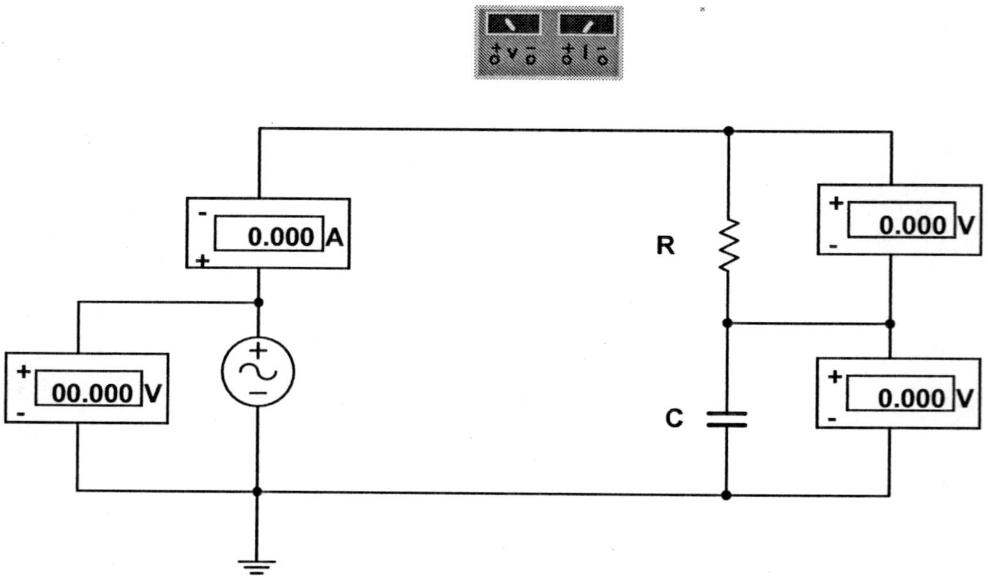

Answers

NAME ______________________________ DATE ______________

23-15. Open the Multisim circuit file CH23 _15 shown.
Use the multimeter, wattmeter, and whatever math calculations may be necessary to complete the table of values for the circuit shown.

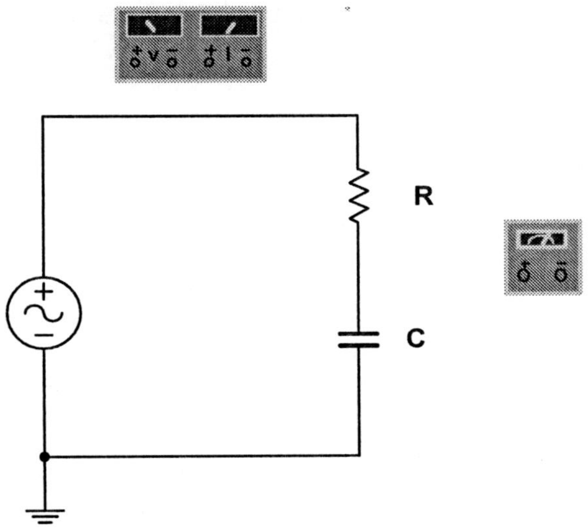

Answers

	E	I	R-X_C-Z	W-VA-VARs	$\angle\theta$	PF
R						
C						
Total						

NAME ____________________ DATE ____________

23-16. Open the Multisim circuit file CH23 _16 shown.
Inputs "A" and "B" of the scope are connected to show the inductor and capacitor voltage waveforms. Turn the simulation on, double click the scope icon and examine the phase shift between the two waveforms displayed. What is the phase relationship between the inductor and capacitor voltages?

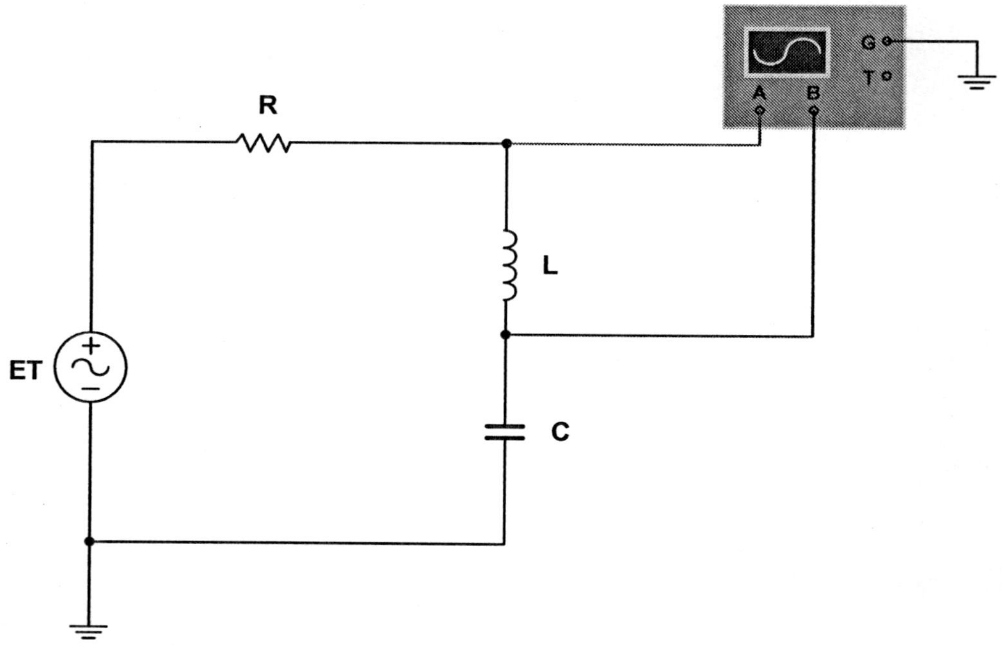

Answer

NAME ______________________ DATE ______________

23-17. Open the Multisim circuit file CH23 _17 shown.

(a) Turn the simulation on and record the value of the voltage readings across the resistor, inductor and capacitor.

(b) Calculate the approximate value of the applied voltage based on the voltmeter readings.

(c) Connect the multimeter into the circuit and measure the actual applied voltage value.

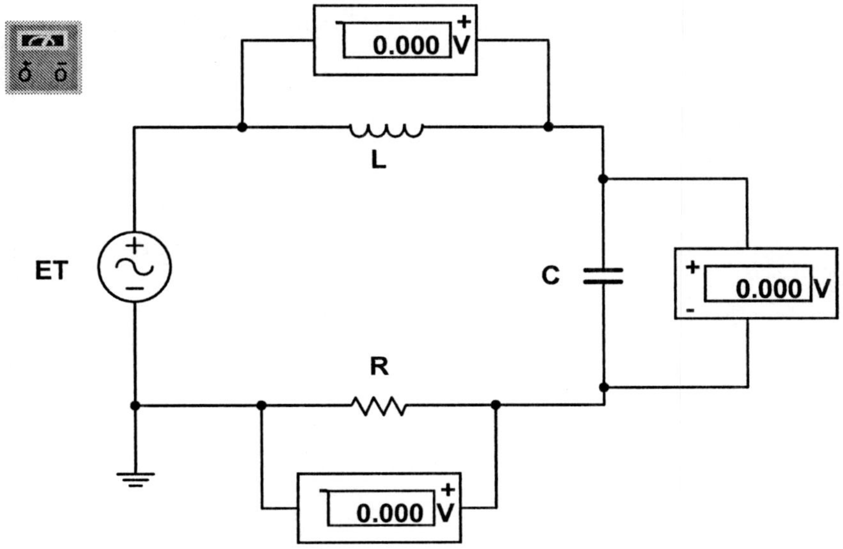

Answers

NAME ____________________ DATE ____________

23-18. Open the Multisim circuit file CH23 _18 shown.

(a) Calculate the value of the impedance.

(b) Calculate the value of the current.

(c) Connect the multimeter into the circuit and measure the actual current value

(d) Calculate the value of the voltage across the inductor, capacitor, and resistor.

(e) Connect the multimeter into the circuit and measure the actual inductor, capacitor, and resistor voltage values.

(f) Calculate the phase angle of the circuit. Double click the scope icon and observe the phase angle of the circuit.

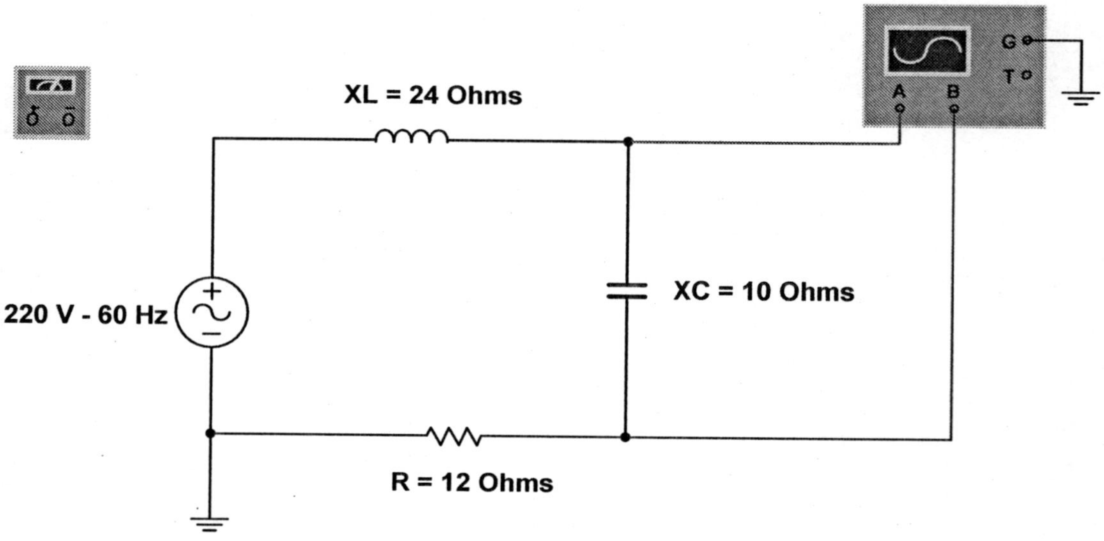

Answers

NAME ______________________________ DATE ____________

23-19. Open the Multisim circuit file CH23 _19 shown and turn the simulation on.

(a) Calculate the true power (W)
(b) Calculate the apparent power (VA)
(c) Calculate the VARs inductive
(d) Calculate the VARs capacitive
(e) Calculate the net VARs
(f) Calculate the power factor (PF)
(g) Connect the wattmeter into the circuit and measure the actual true power and power factor.

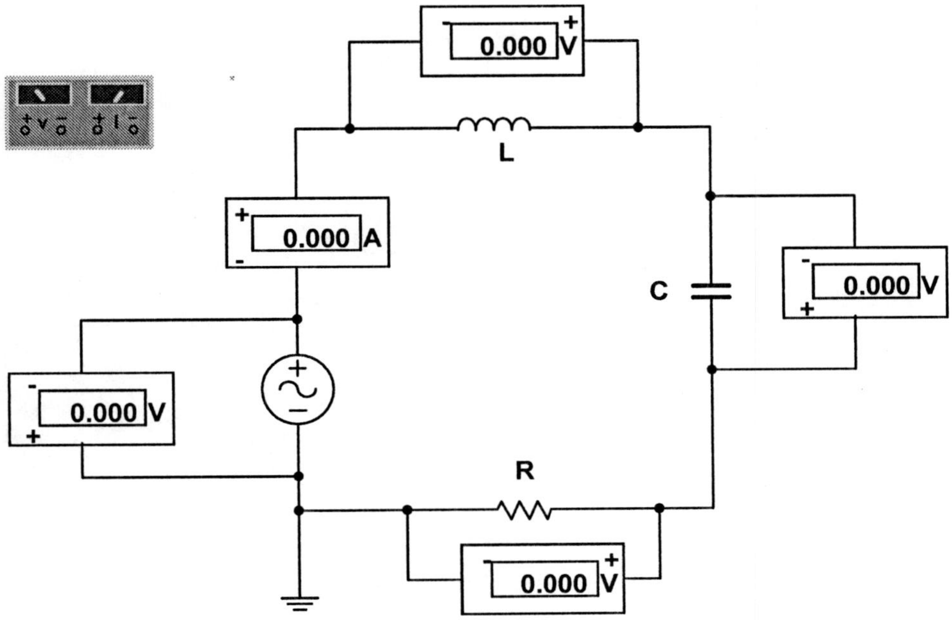

Answers

NAME ______________________ DATE ______________

23-20. Open the Multisim circuit file CH23 _20 shown and turn the simulation on.

(a) Calculate the apparent power

(b) Calculate the true power

(c) Calculate the power factor

(d) Connect the wattmeter into the circuit and measure the true power and power factor.

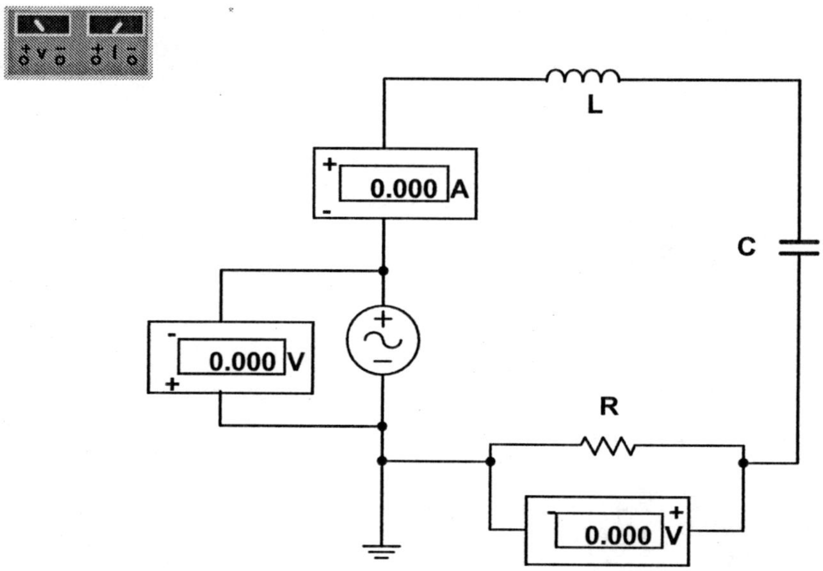

Answers

NAME ______________________________ DATE ______________

23-21. Open the Multisim circuit file CH23 _21 shown.
Use the multimeter, wattmeter, and whatever math calculations may be necessary to complete the table of values for the circuit shown.

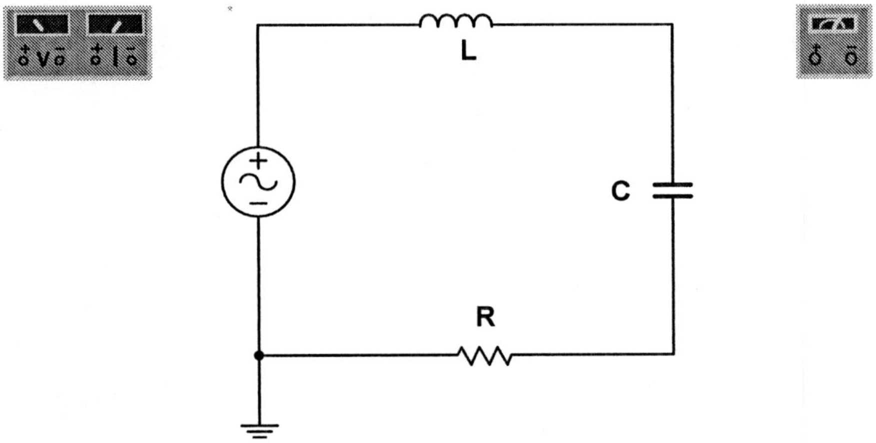

Answers

	E	I	$R/X_L/X_C/Z$	W/ VA /VARs	PF
R					
L					
C					
Total					

NAME ______________________________ DATE ________________

23-22. Open the Multisim circuit file CH23 _22 shown. For resonant RLC circuit, connect the multimeter and wattmeter to measure:

(a) the applied voltage (E_T)
(b) current flow (I)
(c) the voltage across the resistor (E_R)
(d) the voltage across the inductor (E_L)
(e) the voltage across the capacitor (E_C)
(f) the voltage across the series-connected inductor and capacitor (E_L-to-E_C)
(g) true power in watts
(h) power factor

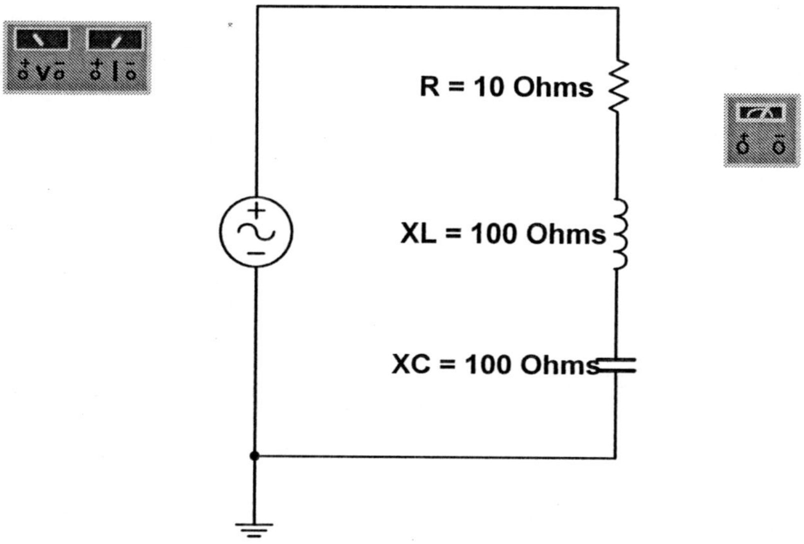

Answers

NAME ________________________ DATE ______________

23-23. Open the Multisim circuit file CH23 _23 shown.

(a) What two voltage waveforms is the scope connected to view? Turn the simulation on and view the two waveforms. What is the phase relationship between the two voltages?

(b) Determine the value of the impedance?

(c) Calculate the value of the current flow.

(d) Connect the multimeter and measure the value of the current flow.

(e) Determine the value of the voltage across the resistor.

(f) Connect the multimeter and measure the voltage across the resistor.

(g) Determine the value of the voltage across the inductor.

(h) Connect the multimeter and measure the voltage across the inductor.

(i) Determine the value of the voltage across the capacitor.

(j) Connect the multimeter and measure the voltage across the capacitor.

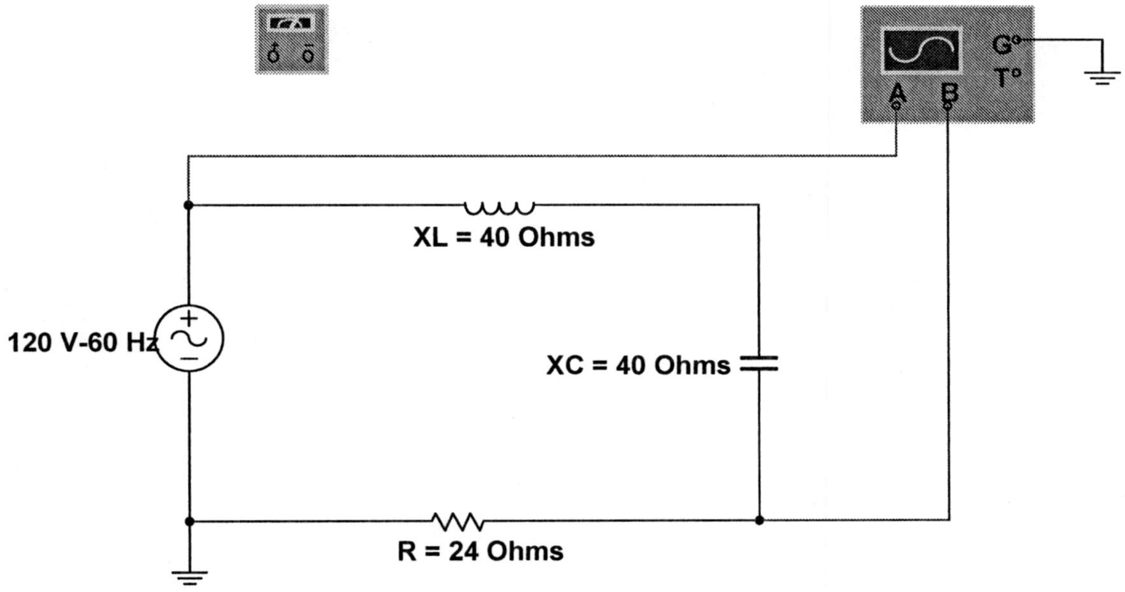

Answers

NAME ________________________________ DATE ________________

23-24. Open the Multisim circuit file CH23 _24 shown.

(a) Calculate the resonant frequency for the series RLC circuit based on the given values of L and C.

(b) When the applied 240-volts is set to the resonant frequency determine the value of the current flow and the voltage across R, L, and C.

(c) Double click on the applied voltage source and set the 240-volts to the resonant frequency. Connect the multimeter and measure the value of the current flow and the voltage across R, L, and C.

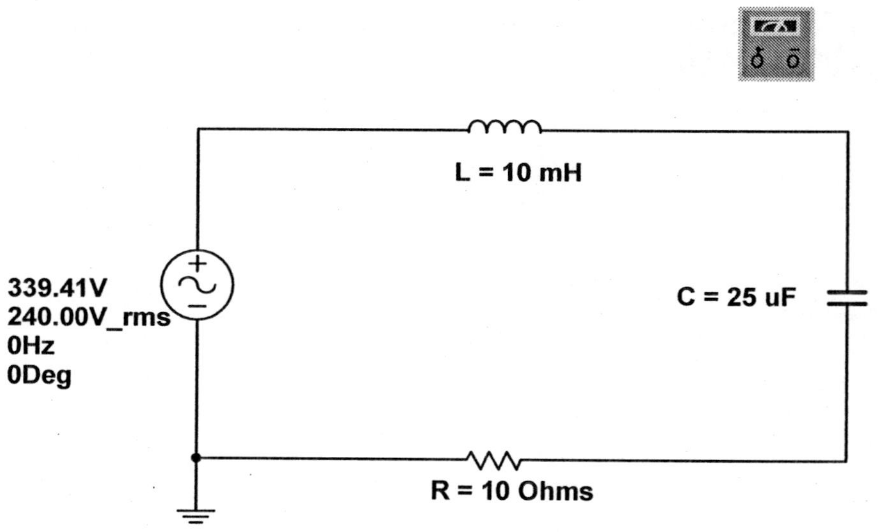

Answers

NAME ______________________________ DATE ________________

23-25. Open the Multisim circuit file CH23 _25 shown.

(a) For the circuit set for the given resonant frequency, turn the simulation on and record the value of the current flow and the voltage drops across R, L, and C.

(b) Increase the frequency of the voltage source above resonance to 200 Hz and answer the following:

(i) does the original current increase or decrease? Why?

(ii) does the original E_R increase or decrease? Why?

(iii) does the original E_L increase or decrease? Why?

(iv) does the original E_C increase or decrease? Why?

(v) Is the circuit now inductive or capacitive in nature? Why?

(c) Decrease the frequency of the voltage source below resonance to 100 Hz and answer the following:

(i) does the original current increase or decrease? Why?

(ii) does the original E_R increase or decrease? Why?

(iii) does the original E_L increase or decrease? Why?

(iv) does the original E_C increase or decrease? Why?

(v) Is the circuit now inductive or capacitive in nature? Why?

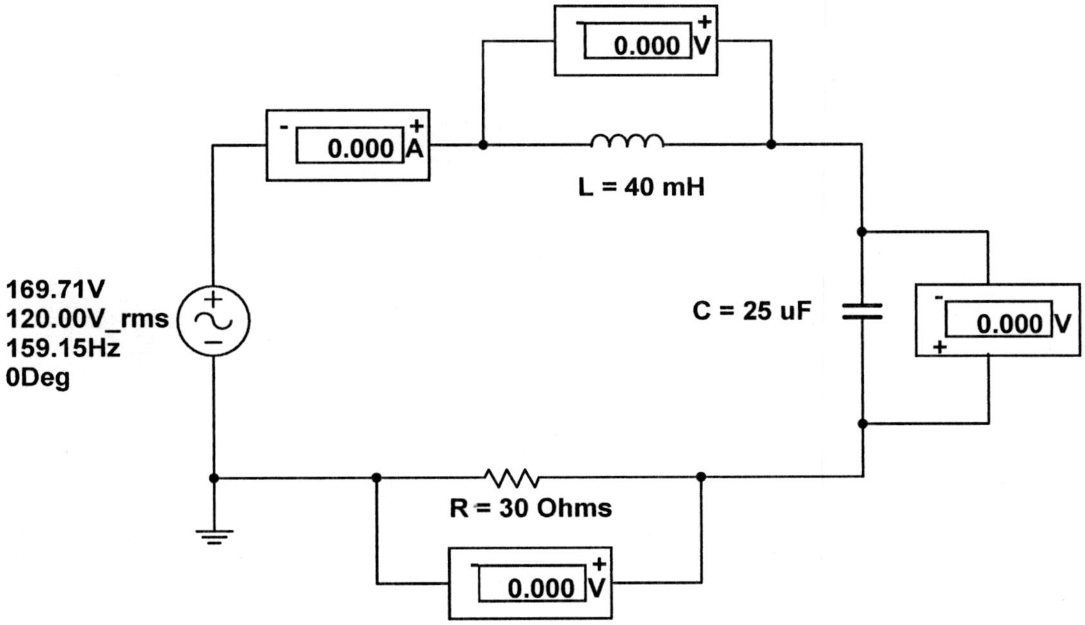

Answers

NAME ____________________ DATE ____________

23-26. Open the Multisim circuit file CH23 _26 shown.

(a) Calculate the value of the cut-off frequency for the low-pass filter circuit shown.

(b) Turn the simulation on and operate the circuit with the input voltage fixed at 10 VAC (rms).

(i) Record the value of the output voltage with the input set to 10 Hz.

(ii) Repeat with the input set to 100 Hz

(iii) Repeat with the input set to 1000 Hz

(c) A Bode plot is a graph of a circuit's frequency response. The Bode plotter connected to the input and output of this circuit will generate a range of frequencies over a specified band and plots a graph of the circuit's frequency response. Double click on the Bode plotter icon and examine the frequency response curve. Record the approximate frequency at which the response curve starts its drop in amplitude.

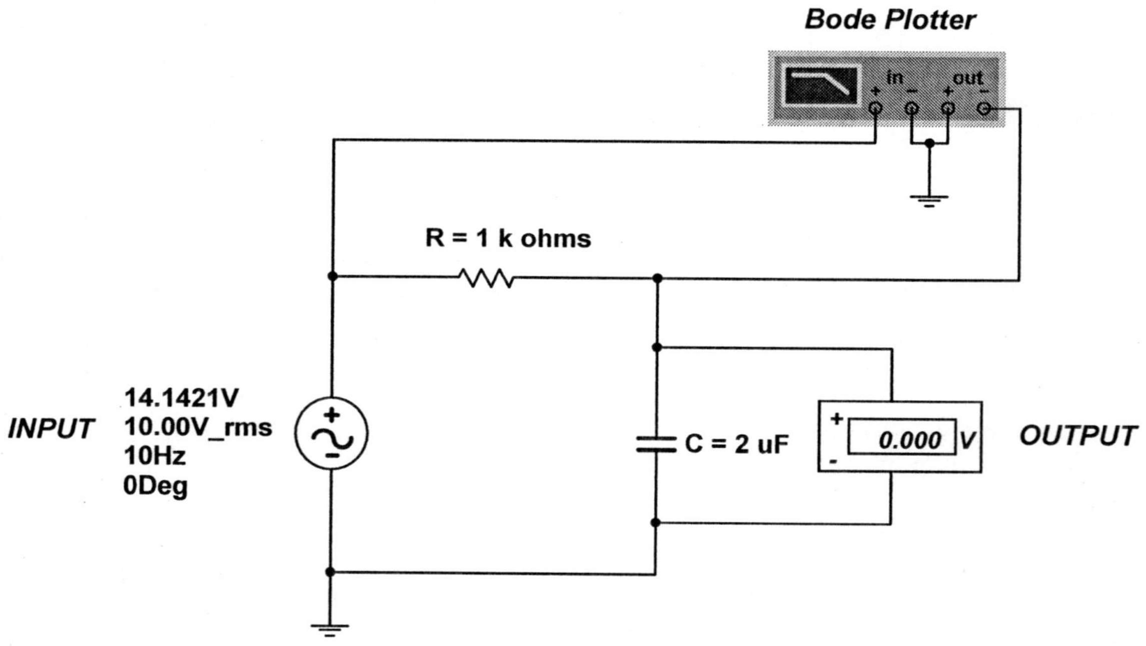

Answers

NAME ______________________________ DATE ______________

23-27. Open the Multisim circuit file CH23 _27 shown.

(a) Calculate the value of the cut-off frequency for the high-pass filter circuit shown.

(b) Turn the simulation on and operate the circuit with the input voltage fixed at 10 VAC (rms).

(i) Record the value of the output voltage with the input set to 10 Hz.

(ii) Repeat with the input set to 100 Hz

(iii) Repeat with the input set to 1000 Hz

(c) Double click on the Bode plotter icon and examine the frequency response curve. Record the approximate frequency at which the response curve starts its drop in amplitude.

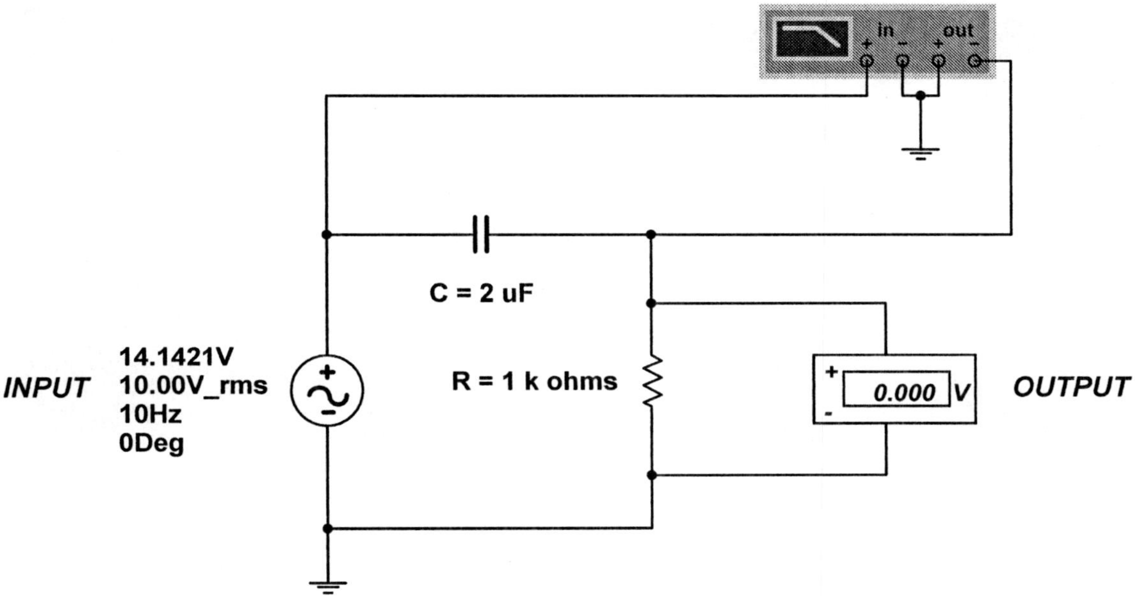

Answers

NAME ______________________________ DATE ____________________

23-28. Open the Multisim circuit file CH23 _28 shown.

(a) Calculate the value of the resonant frequency for the series resonant RLC band-pass circuit shown.

(b) Turn the simulation on and operate the circuit with the input voltage fixed at 10 VAC (rms).

(i) Record the value of the output voltage with the input set to 4 kHz.

(ii) Repeat with the input set to 7 kHz

(iii) Repeat with the input set to 10 kHz

(c) Examine the Q factor of the frequency response curve produced by the Bode plotter. Decrease the value of R to 10 Ω. In what way is the Q of the circuit changed? Why?

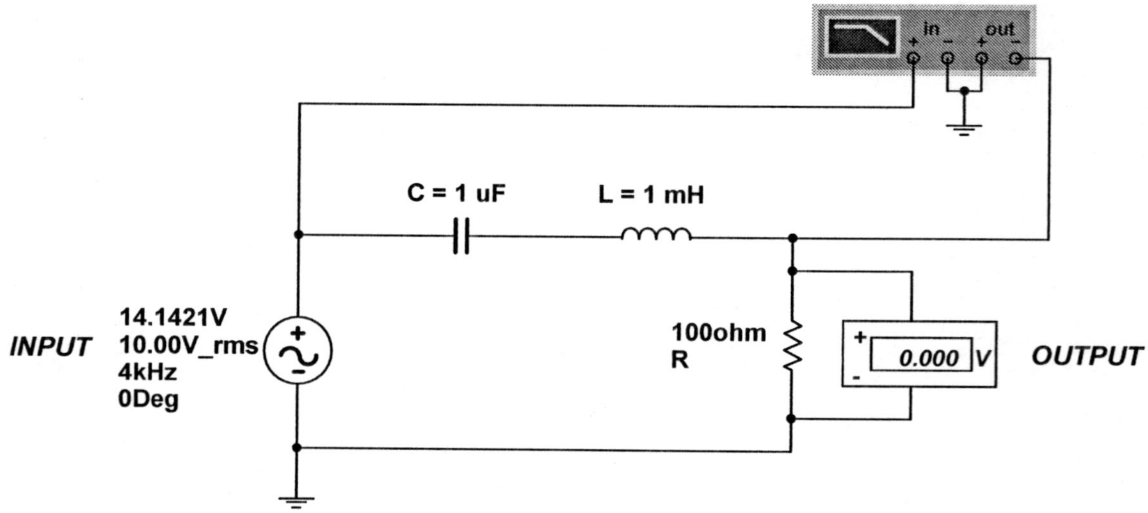

Answers

NAME ______________________ DATE ______________

UNIT 24

RESISTIVE, INDUCTIVE, CAPACITIVE, (RLC) CIRCUITS IN PARALLEL

Multisim Simulation Assignments

24-1. Open the Multisim circuit file CH24 _01 shown and turn the simulation on.

(a) Connect the multimeter and record the value of E_T, E_R, and E_L.

(b) Connect the multimeter and record the value of I_T, I_R, and I_L.

(c) Based on your readings, calculate the impedance (Z) of the circuit

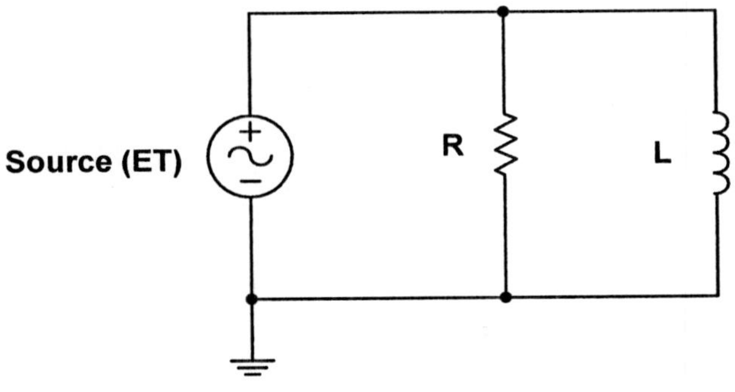

Answers

NAME ______________________________ DATE ________________

24-2. Open the Multisim circuit file CH24 _02 shown.

(a) Calculate the value of I_R.

(b) Connect the multimeter and record the value of I_R.

(c) Calculate the value of I_L.

(d) Connect the multimeter and record the value of I_L.

(e) Calculate the value of I_T.

(f) Connect the multimeter and record the value of I_T.

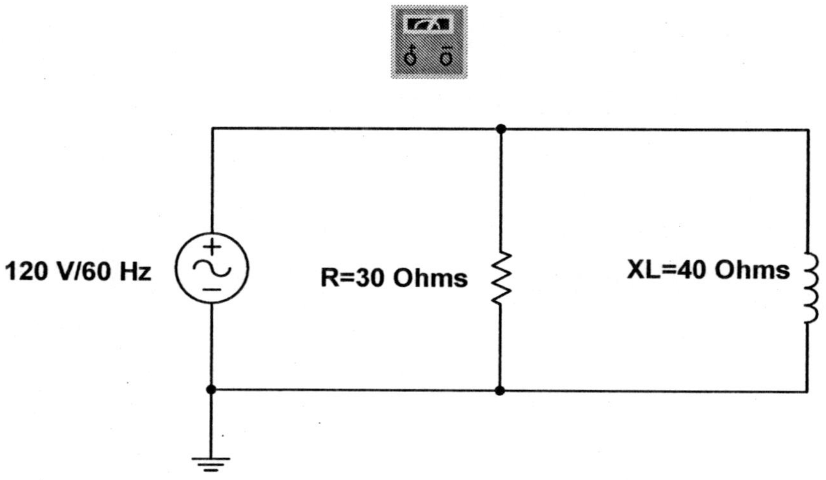

Answers

NAME ________________________________ DATE ________________

24-3. Open the Multisim circuit file CH24 _03 shown.

(a) Calculate the impedance based on the given R and X_L values.
(b) Calculate the total current based on the calculated impedance value.
(c) Connect the multimeter and record the value of I_T.
(d) Calculate the current through R.
(e) Connect the multimeter and record the value of I_R.
(f) Calculate the current through L
(g) Connect the multimeter and record the value of I_L.

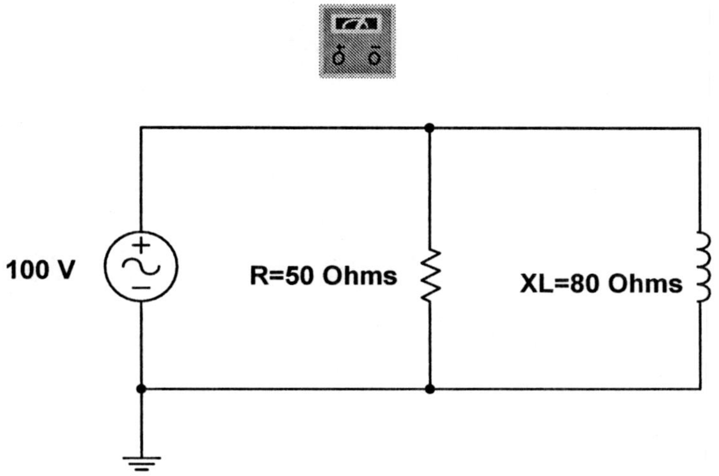

Answers

NAME ______________________________ DATE ________________

24-4. Open the Multisim circuit file CH24 _04 shown.

(a) Record the value of the applied voltage (E), total current (I_T), current through the resistor (I_R), and the current through the inductor (I_L).

(b) Calculate the value of the apparent power.

(c) Calculate the value of the inductive reactive power.

(d) Calculate the value of the true power.

(e) Connect the wattmeter into the circuit and record the wattage reading.

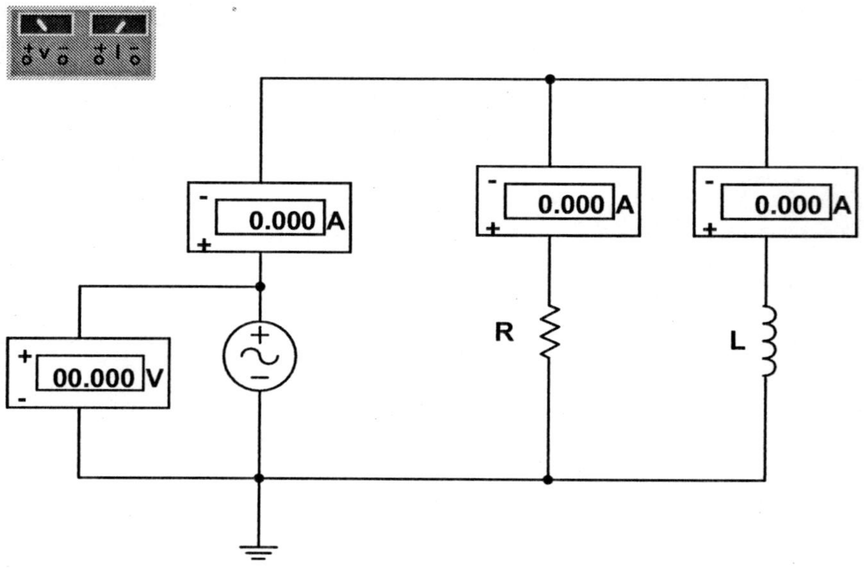

Answers

NAME ____ DATE ____________________

24-5. Open the Multisim circuit file CH24 _05 shown.
Use the multimeter and whatever other math calculations may be necessary to complete the table of values for the circuit shown.

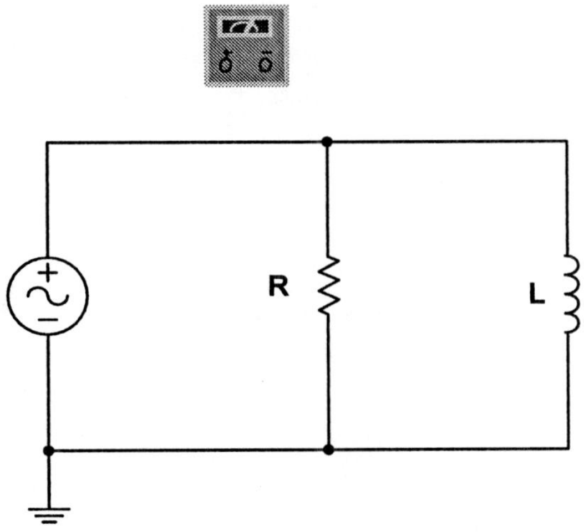

Answers

	E	I	R /X_L /Z	W/VA/VARs	∠θ	PF
R						
L						
Total						

NAME ______________________ DATE ______________

24-6. Open the Multisim circuit file CH24 _06 shown.
Complete the table for all given and unknown quantities for the parallel RL circuit shown. Use the multimeter to check as many of your calculations as possible.

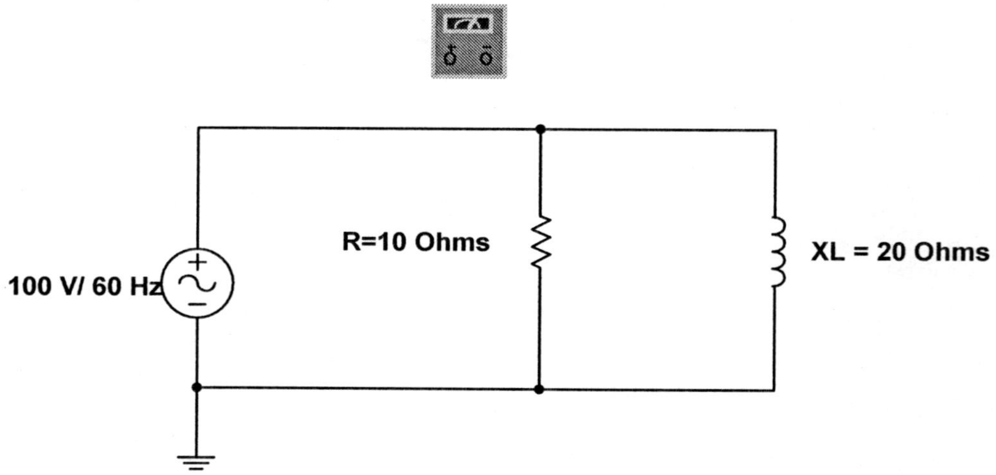

Answers

	E	I	R /X_L /Z	W/VA/VARs
R				
L				
Total				

NAME ________________________________ DATE ________________

24-7. Open the Multisim circuit file CH24 _07 shown.

(a) Calculate the value of I_R.

(b) Connect the multimeter and record the value of I_R.

(c) Calculate the value of I_C.

(d) Connect the multimeter and record the value of I_C.

(e) Calculate the value of I_T.

(f) Connect the multimeter and record the value of I_T.

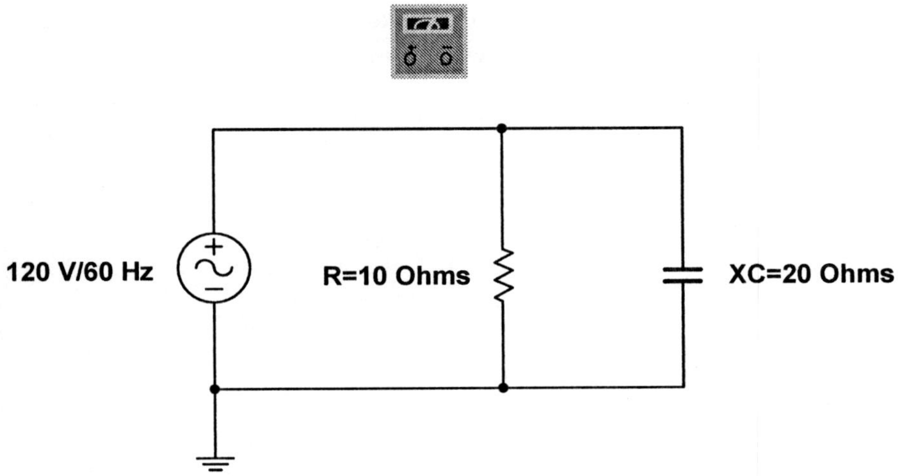

Answers

NAME ______________________________ DATE ________________

24-8. Open the Multisim circuit file CH24 _08 shown and turn the simulation on.

(a) Record the value of the applied voltage (E), total current (I_T), current through the resistor (I_R), and the current through the capacitor (I_C).

(b) Calculate the value of the apparent power.

(c) Calculate the value of the inductive reactive power.

(d) Calculate the value of the true power.

(e) Calculate the value of the power factor

(f) Connect the wattmeter into the circuit and record the wattage and power factor readings.

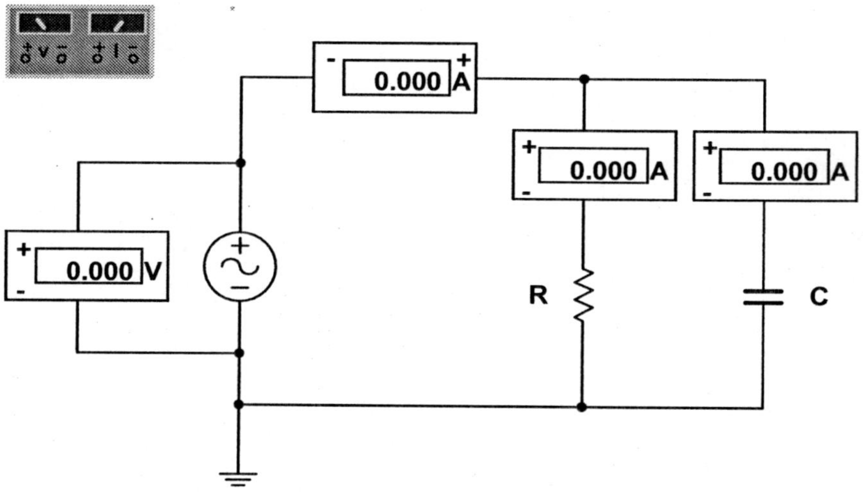

Answers

NAME ________________________________ DATE ________________

24-9. Open the Multisim circuit file CH24 _09 shown.
Complete the table for all given and unknown quantities for the parallel RL circuit shown. Use the multimeter and wattmeter to check as many of your calculations as possible.

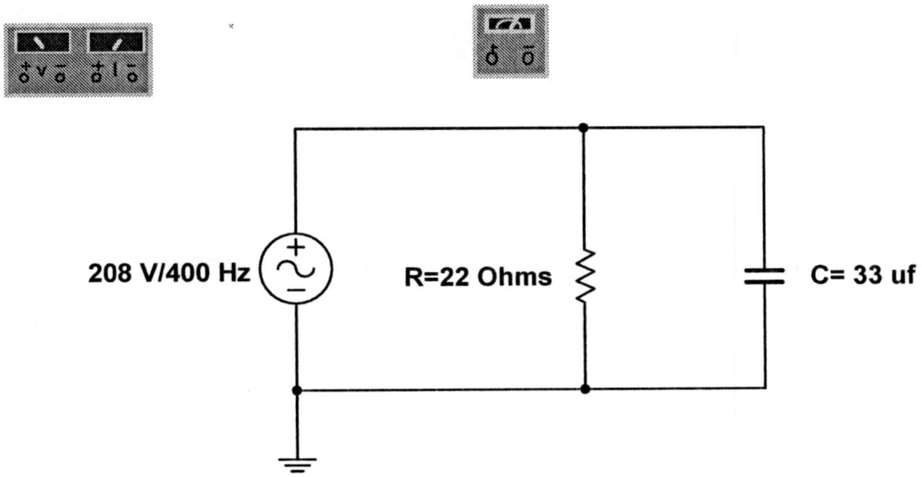

Answers

	E	I	C	R /X_C /Z	W/VA/VARs	∠θ	PF
R							
C							
Total							

NAME ________________________ DATE ______________

24-10. Open the Multisim circuit file CH24 _10 shown and turn the simulation on.

(a) Double click on the wattmeter and record the value of the wattage and power factor readings.

(b) Calculate the value of the current flow through the resistor. Check your answer by measuring the resistor current.

(c) Calculate the resistance value of the resistor. Check your answer by measuring the resistance value.

(d) Calculate the capacitive reactance of the capacitor.

(e) Calculate the value of the current flow through the capacitor. Check your answer by measuring the capacitor current.

(f) Calculate the value of the line current. Check your answer by measuring the line current.

(g) Calculate the value of the apparent power.

(h) Calculate the % PF.

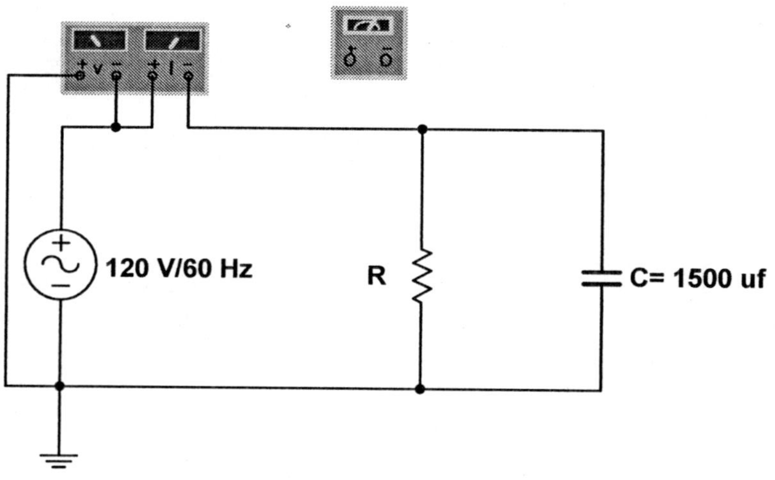

Answers

NAME ______________________________ DATE ______________

24-11. Open the Multisim circuit file CH24 _11 shown and turn the simulation on.
Complete the table for all given and unknown quantities for the parallel RL circuit shown. Use the multimeter and wattmeter to check as many of your calculations as possible.

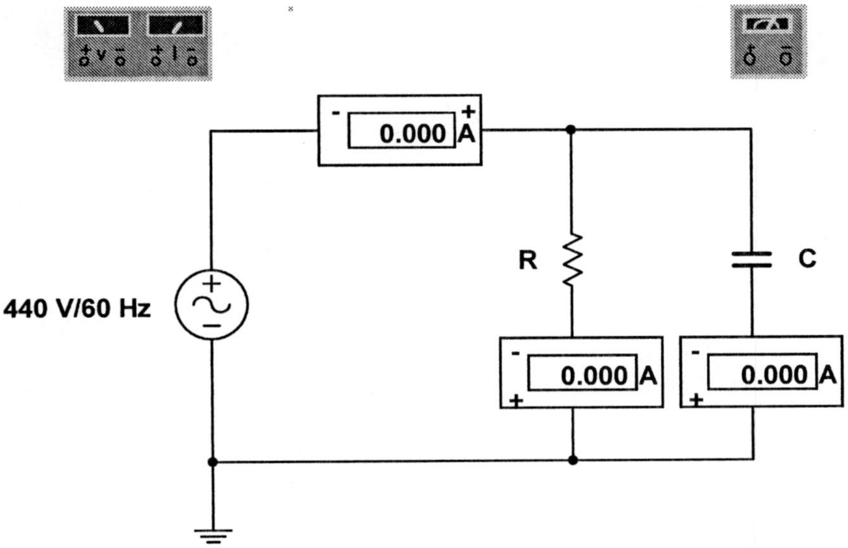

Answers

	E	I	C	R /X_C /Z	W/VA/VARs	∠θ	PF
R							
C							
Total							

NAME ______________________________ DATE ____________________

24-12. Open the Multisim circuit file CH24 _12 shown and turn the simulation on.

(a) Record the value of the current flow through the inductor and capacitor.

(b) Calculate the value of the line current.

(c) Connect the multimeter into the circuit and record the value of the line current.

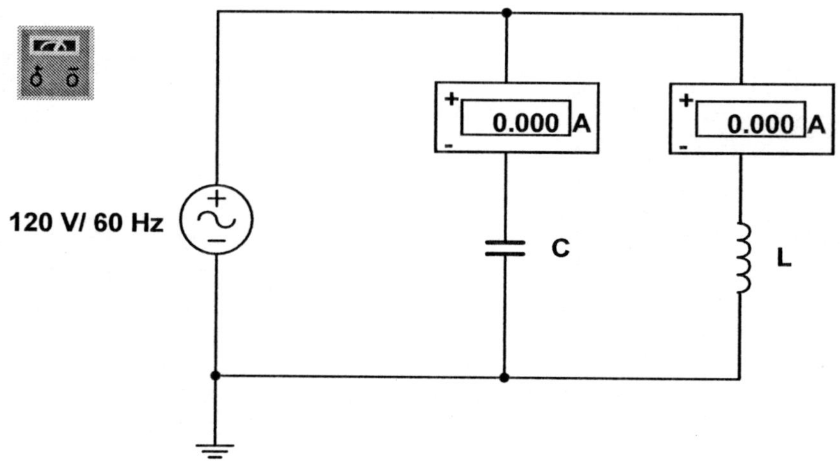

Answers

NAME ______________________________ DATE ________________

24-13. Open the Multisim circuit file CH24 _13 shown.

(a) Calculate the impedance of the circuit based on the reactance values.

(b) Calculate the line current flow based on your impedance calculation.

(c) Connect the multimeter into the circuit and record the value of the line current.

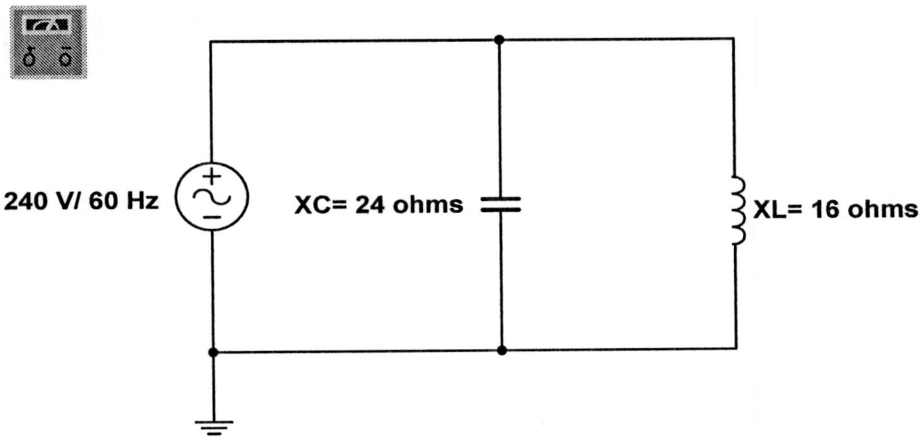

Answers

NAME ______________________________ DATE ________________

24-14. Open the Multisim circuit file CH24 _14 shown and turn the simulation on.

(a) Record the value of the current flow through the inductor, capacitor, and line.

(b) Determine the apparent power.

(c) Determine the reactive power.

(d) Determine the true power.

(e) Determine the power factor

(c) Connect the wattmeter into the circuit and record the value of the watts and power factor.

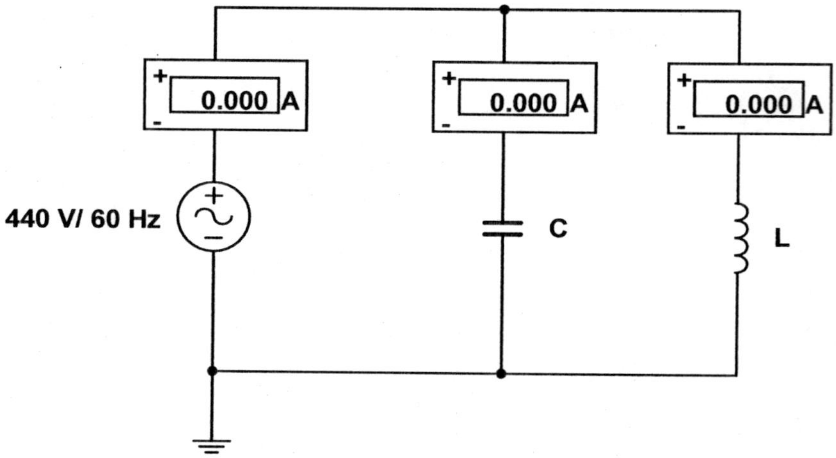

Answers

NAME ____________________ DATE ____________

24-15. Open the Multisim circuit file CH24 _15 shown and turn the simulation on. Connect the multimeter and record the value of:

(a) The applied voltage and the voltage across R, L, and C. How do they compare in value ? Why?

(b) The current flow through the resistor.

(c) The current flow through the inductor.

(d) The current flow through the capacitor

(e) The reactive current (I_X) between points "A" and "B".

(f) The total line current.

(g) Is the circuit inductive or capacitive in nature? Why?

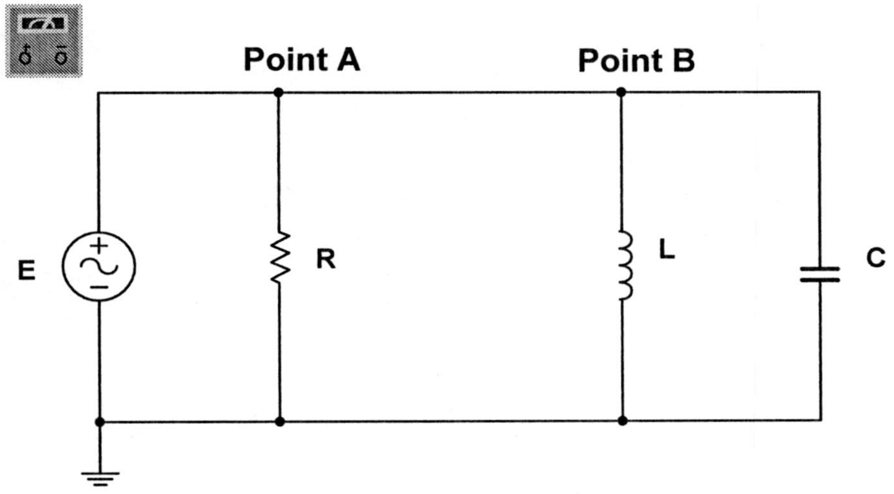

Answers

NAME ______________________________ DATE ____________________

24-16. Open the Multisim circuit file CH24 _16 shown and turn the simulation on.

(a) Record the value of the current flow through the resistor, inductor, and capacitor.

(b) Calculate the value of the current flow between points "A" and "B" of the circuit.

(c) Calculate the value of the total line current.

(d) Connect the multimeter and record the value of the current flow between points "A" and "B".

(e) Connect the multimeter and record the value of the line current flow.

(f) Is the circuit inductive or capacitive in nature? Why?

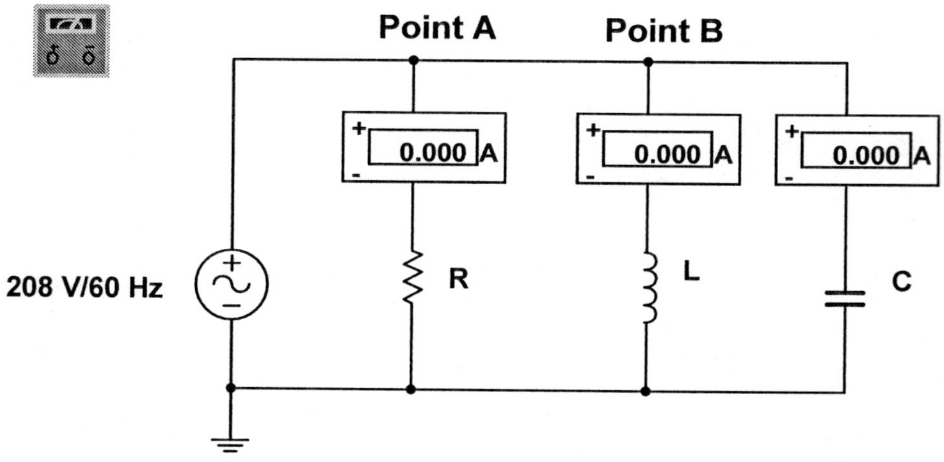

Answers

NAME ________________________ DATE ______________

24-17. Open the Multisim circuit file CH24 _1 shown.

(a) Calculate the value of the current flow through the resistor. Check your answer by measuring the resistor current.

(b) Calculate the value of the current flow through the inductor. Check your answer by measuring the inductor current.

(c) Calculate the value of the current flow through the capacitor. Check your answer by measuring the capacitor current.

(d) Calculate the value of the net reactive current flow. Check your answer by measuring the net reactive current.

(e) Calculate the value of the total line current flow. Check your answer by measuring the line current.

(f) Calculate the value of the circuit impedance.

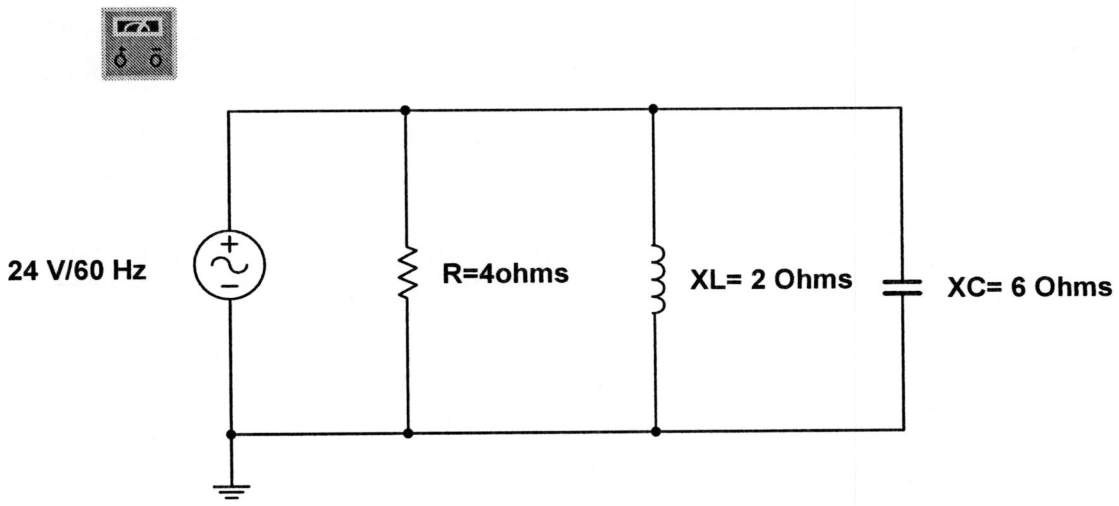

Answers

NAME ____________________ DATE ____________

24-18. Open the Multisim circuit file CH24 _18 shown and turn the simulation on.

(a) Record the value of I_T, I_R, I_X, I_L and I_C.

(b) Calculate the value of the true power.

(c) Calculate the value of the reactive power.

(d) Calculate the value of the apparent power.

(e) Calculate the power factor

(f) Connect the wattmeter into the circuit and record the value of the watts and power factor.

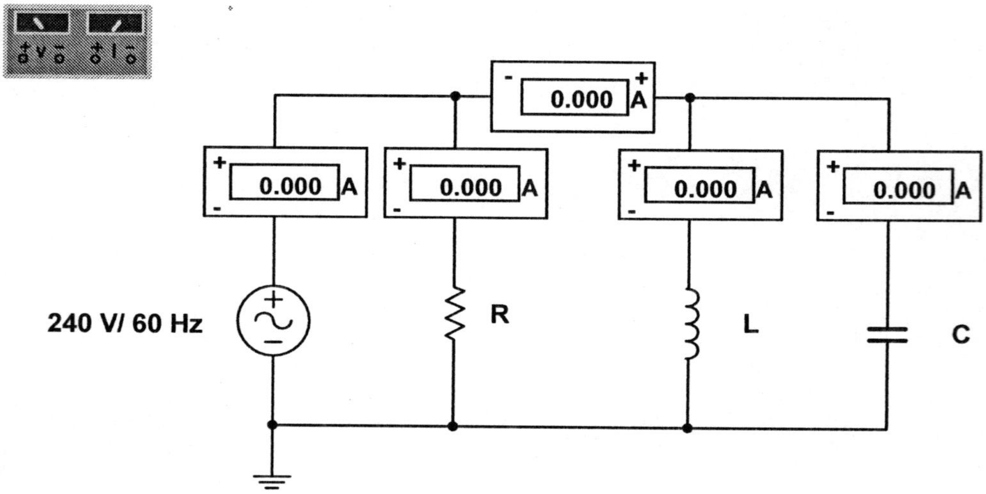

Answers

NAME ______________________ DATE ______________

24-19. Open the Multisim circuit file CH24 _19 shown.

(a) Calculate the value of the current flow through the resistor. Check your answer by measuring the resistor current.

(b) Calculate the value of the current flow through the inductor. Check your answer by measuring the inductor current.

(c) Calculate the value of the current flow through the capacitor. Check your answer by measuring the capacitor current.

(d) Calculate the value of the net reactive current flow. Check your answer by measuring the net reactive current.

(e) Calculate the value of the total line current flow. Check your answer by measuring the line current.

(f) Calculate the value of the circuit impedance.

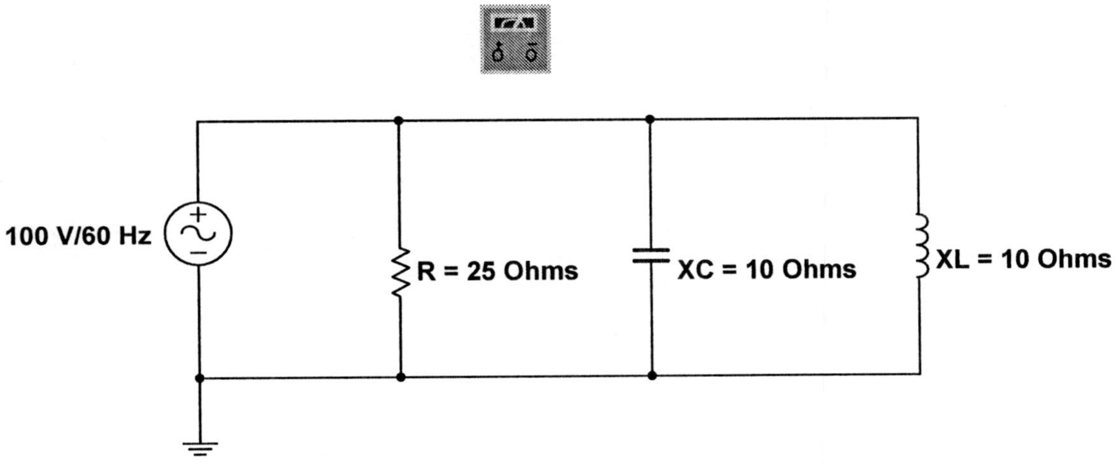

Answers

NAME ______________________________ DATE ______________

24-20. Open the Multisim circuit file CH24 _20 shown and turn the simulation on.

(a) Record the value of the true power and power factor reading of the wattmeter.

(b) Calculate the value of the line current.

(c) Connect the multimeter into the circuit and record the value of the line current.

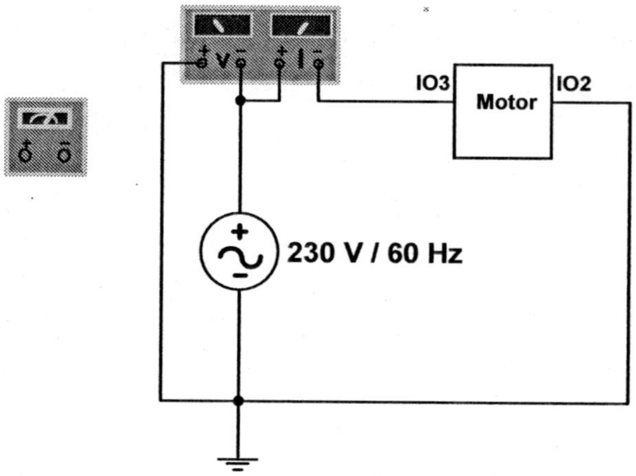

Answers

NAME ______________________ DATE ______________

24-21. Open the Multisim circuit file CH24 _21 shown and turn the simulation on.

(a) Record the value of the applied voltage and line current flow.
(b) Calculate the kVA power of the circuit.
(c) Calculate the power factor of the circuit.
(d) Connect the wattmeter into the circuit and record the wattage and power factor readings.

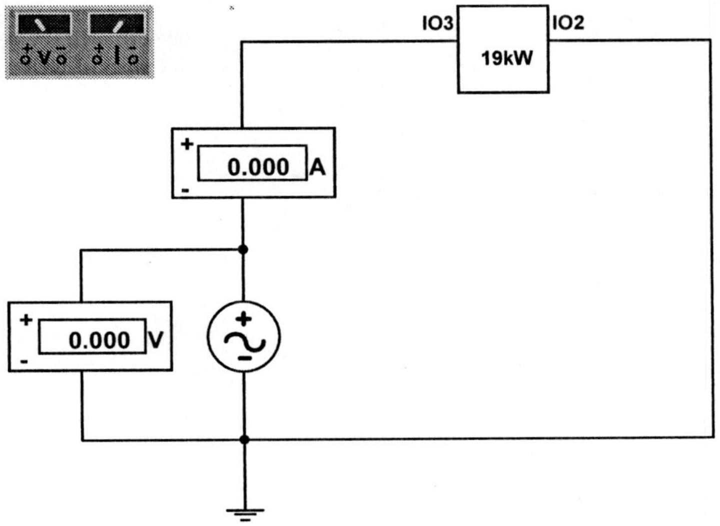

Answers

NAME ________________________________ DATE __________________

24-22. Open the Multisim circuit file CH24 _22 shown and turn the simulation on.

(a) Connect the multimeter and wattmeter into the circuit and record the value of the (i) line current flow (ii) motor current flow (iii) capacitor current flow (iv) true power (v) power factor

(b) Repeat part (a) with the capacitor disconnected from the circuit.

(c) Which readings changed? Which readings stayed the same?

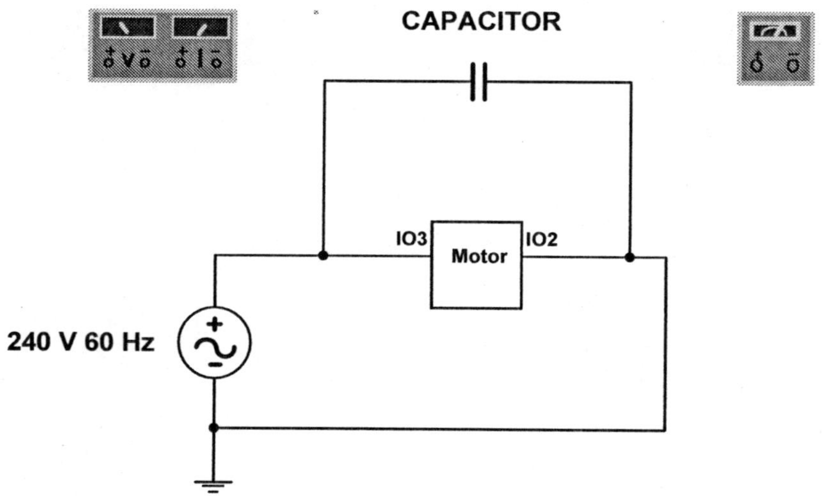

Answers

NAME ________________________________ DATE ________________

24-23. Open the Multisim circuit file CH24 _23 shown and turn the simulation on.

(a) Record the value of the wattmeter reading, power factor reading, ammeter reading and voltmeter reading for the simulated motor circuit.

(b) Calculate the value of the apparent power and reactive power.

(c) Calculate the value of the apparent power and reactive power with a capacitor connected to achieve a corrected power factor of 0.90.

(d) Calculate the capacitive VARs required to raise the power factor to 0.90.

(e) Calculate the capacitive reactance of the capacitor required to raise the power factor to 0.90.

(f) Size of capacitor, connected in parallel with the motor, required to raise the power factor to 0.90.

(g) Calculate the amount of reduction in line current with the power factor corrected to 0.90.

(h) Set the capacitor value to that required to correct the power factor to 0.90. Wire the capacitor in parallel with the motor and record the value of the wattmeter reading, power factor reading, ammeter reading and voltmeter reading

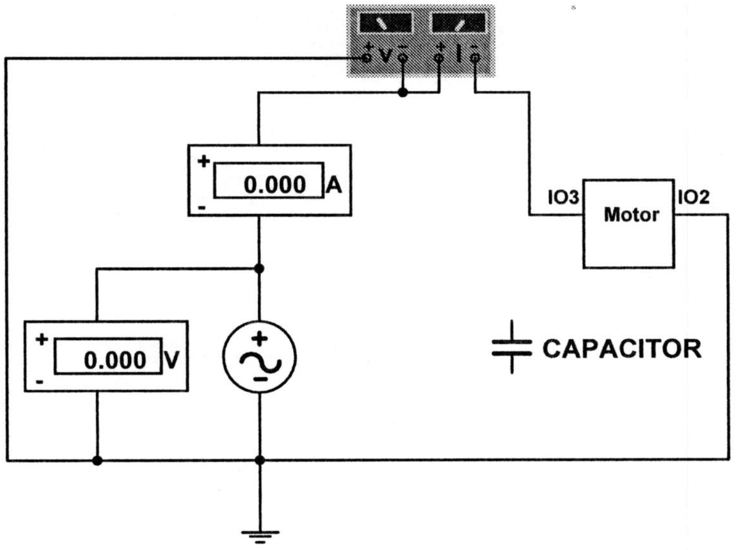

Answers

NAME ______________________________ DATE ________________

24-24. Open the Multisim circuit file CH24 _24 shown.

(a) Calculate the value of the current flow through branch Z_1. Verify your calculations by measuring this current.

(b) Calculate the value of the current flow through branch Z_2. Verify your calculations by measuring this current.

(c) Calculate the value of the current flow through branch Z_3. Verify your calculations by measuring this current.

(d) Calculate the value of the total line current flow. Verify your calculations by measuring this current.

(e) Calculate the total impedance of the circuit.

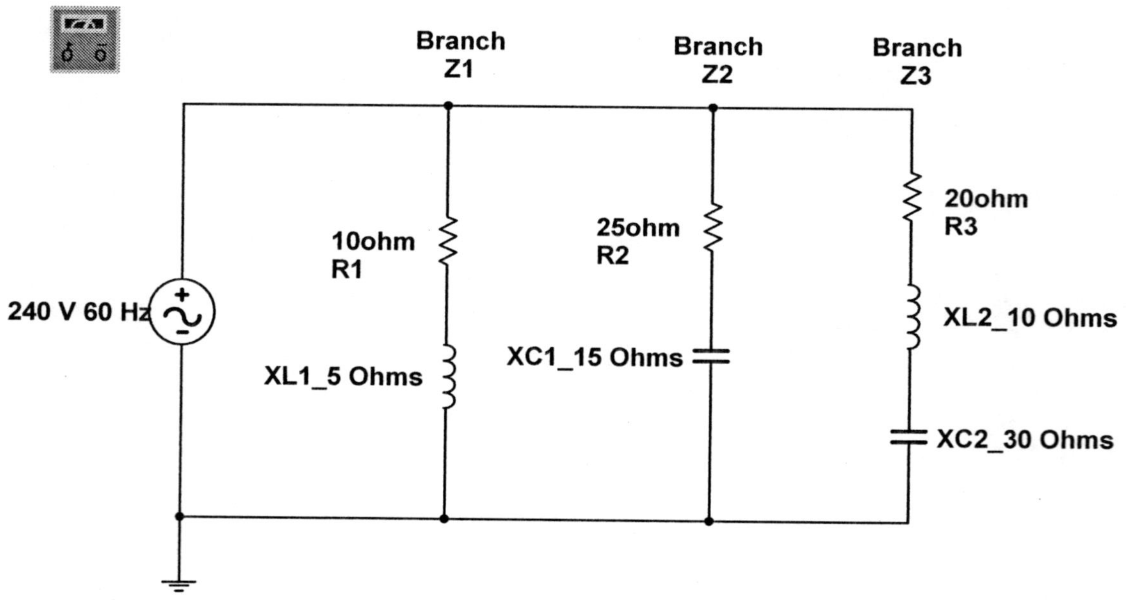

Answers

NAME ______________________________ DATE ________________

24-25. Open the Multisim circuit file CH24 _25 shown and turn the simulation on. Calculate the impedance of the circuit.

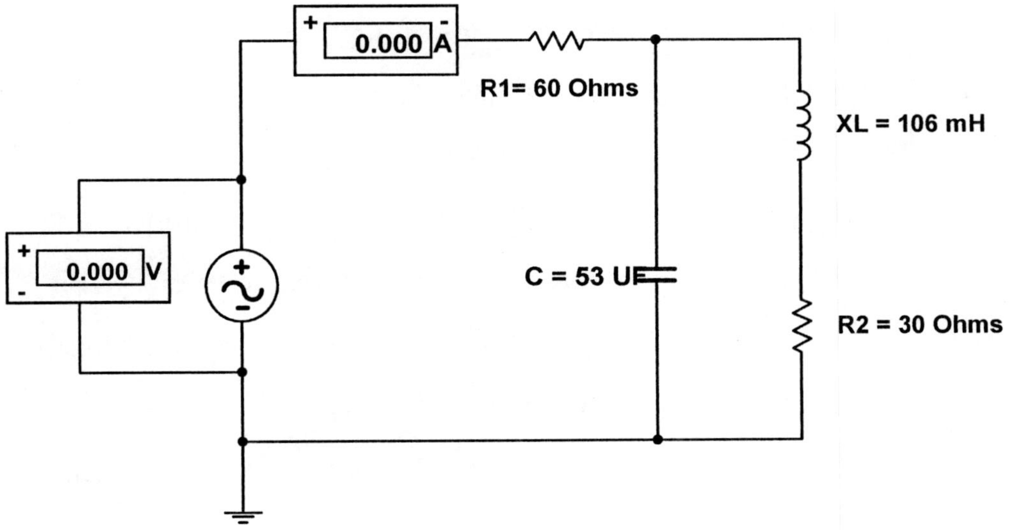

Answers

NAME ______________________________ DATE ________________

24-26. Open the Multisim circuit file CH24 _26 shown and turn the simulation on. Calculate:

(a) impedance of the capacitive branch (Z_1)

(b) impedance of the inductive branch (Z_2)

(c) total impedance of the circuit (Z_T)

(d) voltage drop across R_1 (verify by measurement)

(e) voltage drop across C (verify by measurement)

(f) voltage drop across R_2 (verify by measurement)

(g) voltage drop across L (verify by measurement)

(h) true power of the circuit (Watts) (verify by measurement)

(i) apparent power of the circuit (VA)

(j) power factor of the circuit (PF) (verify by measurement)

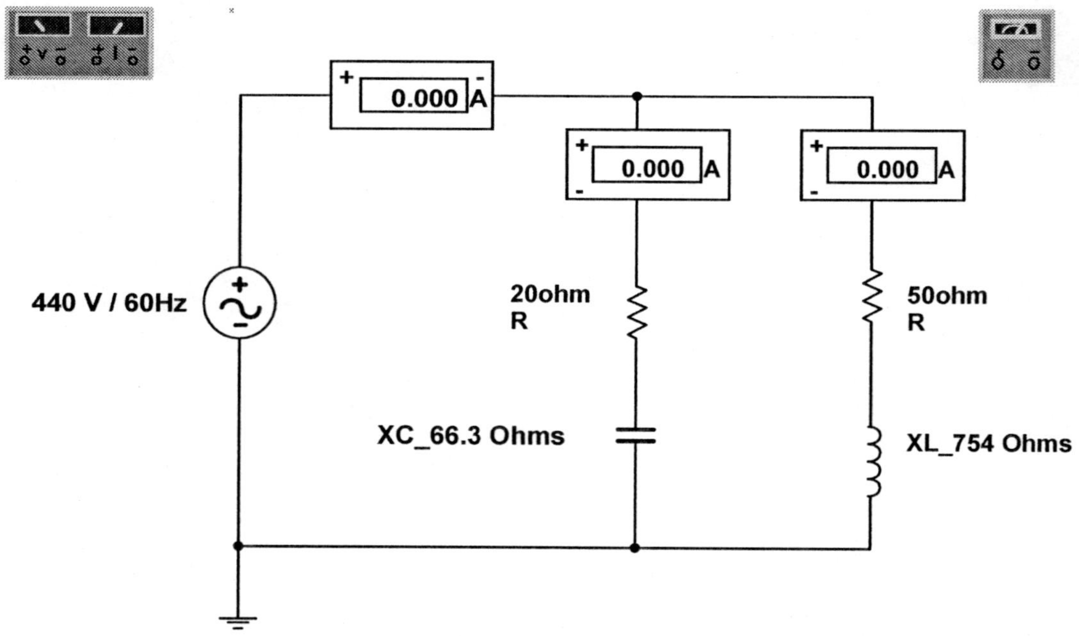

Answers

NAME ______________________________ DATE ________________

24-27. Open the Multisim circuit file CH24 _27 shown. Calculate:

(a) the impedance of the inductive branch

(b) the value of the current through the inductive branch (verify by measurement)

(c) the value of the current through the capacitive branch (verify by measurement)

(d) the value of the line current (verify by measurement)

(e) the value of the voltage drop across the resistor (verify by measurement)

(f) the value of the voltage drop across the inductor (verify by measurement)

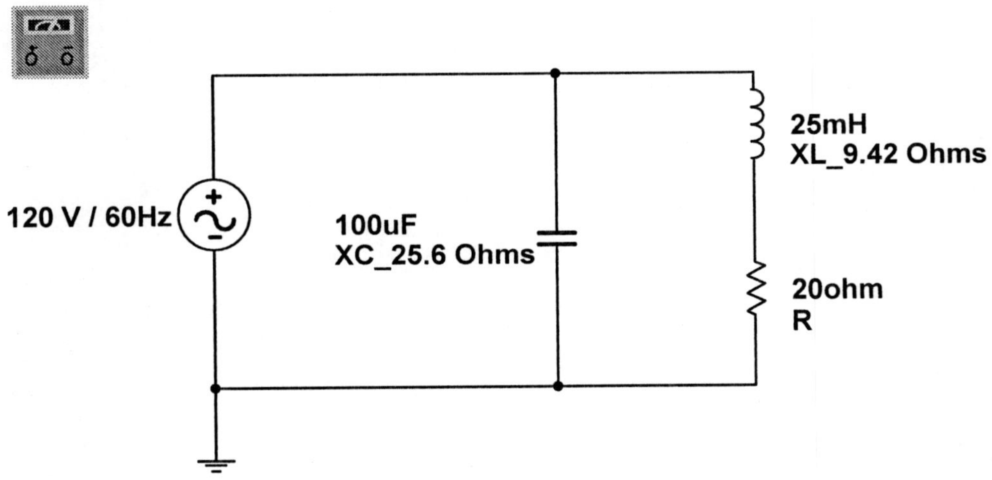

Answers

NAME ____________________________________ DATE ____________________

UNIT 25

TRANSFORMERS

Multisim Simulation Assignments

25-1. Open the Multisim circuit file CH25 _01 shown.

(a) Determine the turns ratio of the transformer

(b) Calculate the value of the secondary voltage.

(c) Connect the multimeter into the circuit and record the value of the secondary voltage.

(d) What type of transformer would this be classified as? Why?

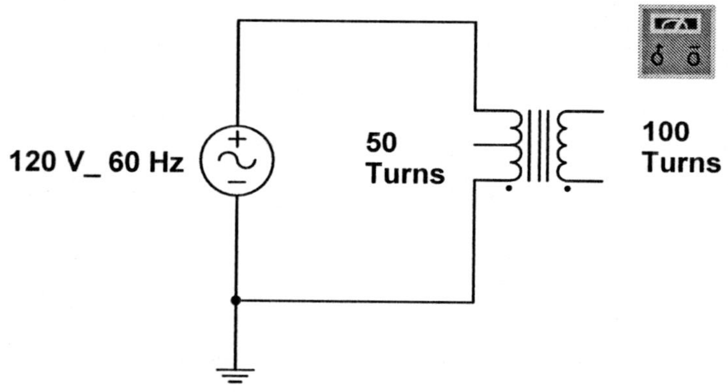

Answers

NAME ______________________________ DATE ________________

25-2. Open the Multisim circuit file CH25 _02 shown.

(a) Determine the turns ratio of the transformer

(b) Calculate the value of the secondary voltage.

(c) Connect the multimeter into the circuit and record the value of the secondary voltage.

(d) What type of transformer would this be classified as? Why?

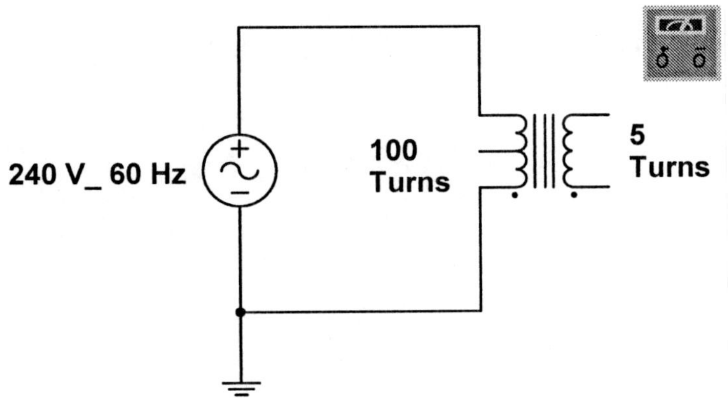

Answers

NAME ____________________ DATE ____________

25-3. Open the Multisim circuit file CH25 _03 shown and turn the simulation on.

(a) Record the value of the primary and secondary voltages.

(b) Calculate the value of the secondary current.

(c) Measure the value of the secondary current.

(d) Calculate the volt-amps of the secondary.

(e) Calculate the volt-amps of the primary.

(f) Calculate the value of the primary current.

(g) Measure the value of the primary current.

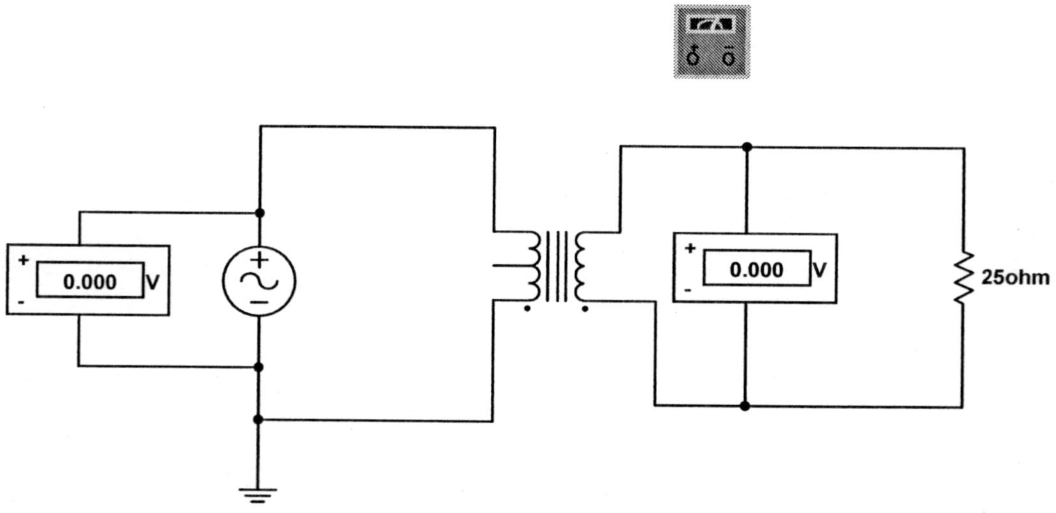

Answers

NAME ______________________________ DATE ________________

25-4. Open the Multisim circuit file CH25 _04 shown and turn the simulation on.

(a) Record the value of E_P, E_S, and I_S.

(b) Calculate the value of the primary current.

(c) Measure the value of the primary current.

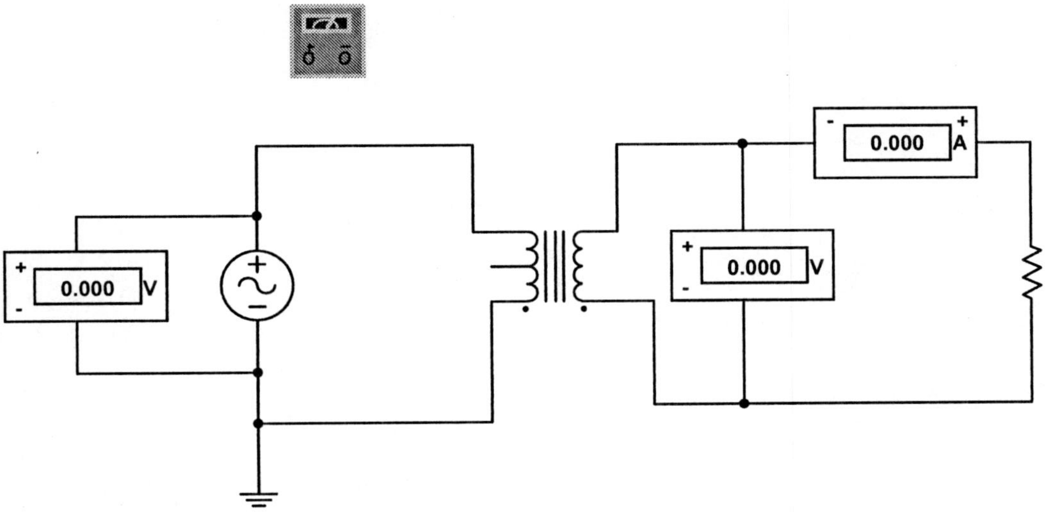

Answers

NAME ______________________ DATE ______________

25-5. Open the Multisim circuit file CH25 _05 shown. This circuit simulates a polarity test being made on the transformer.

(a) Turn the simulation on and record the value of the two voltmeter readings.

(b) What type of polarity is indicated?

(c) What is the value of the voltage across the secondary? (verify by measurement)

(c) What would the secondary lead with the dot be labeled as?

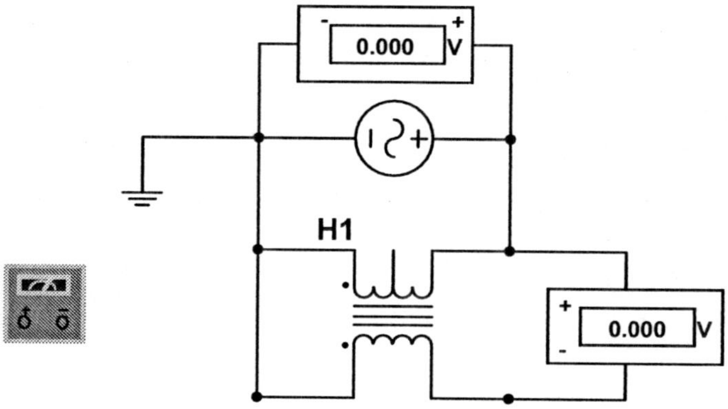

Answers

NAME ______________________________ DATE ______________

25-6. Open the Multisim circuit file CH25 _06 shown.

(a) Complete the wiring of the transformer so that 120-volts will be supplied equally to each of the two separate loads.

(b) Record the value of the voltage across Load 1 and Load 2.

(c) Record the value of the current through Load 1 and Load 2.

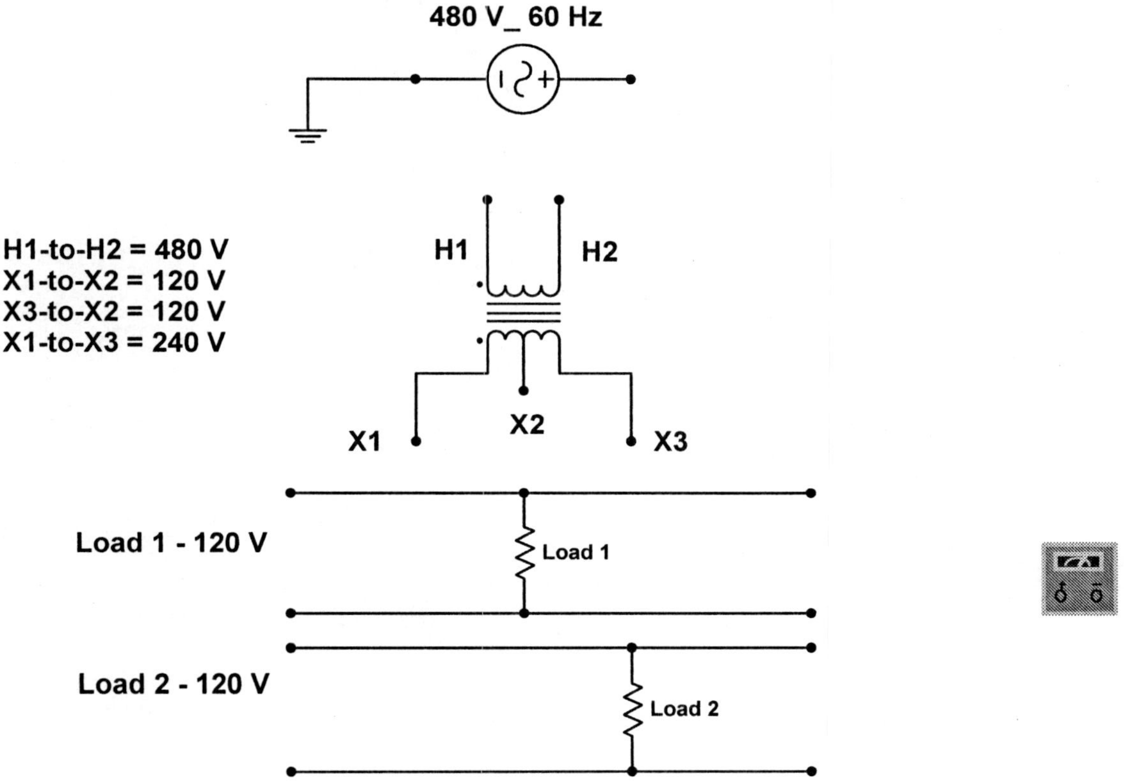

Answers

NAME ____________________ DATE ____________

25-7. Open the Multisim circuit file CH25 _07 shown.

(a) Complete the wiring of the transformer so that both 120-volts and 240-volts are supplied to their respective loads.

(b) Record the value of the voltage across Load 1, Load 2, and Load 3.

(c) Record the value of the current through Load 1, Load 2, and Load 3.

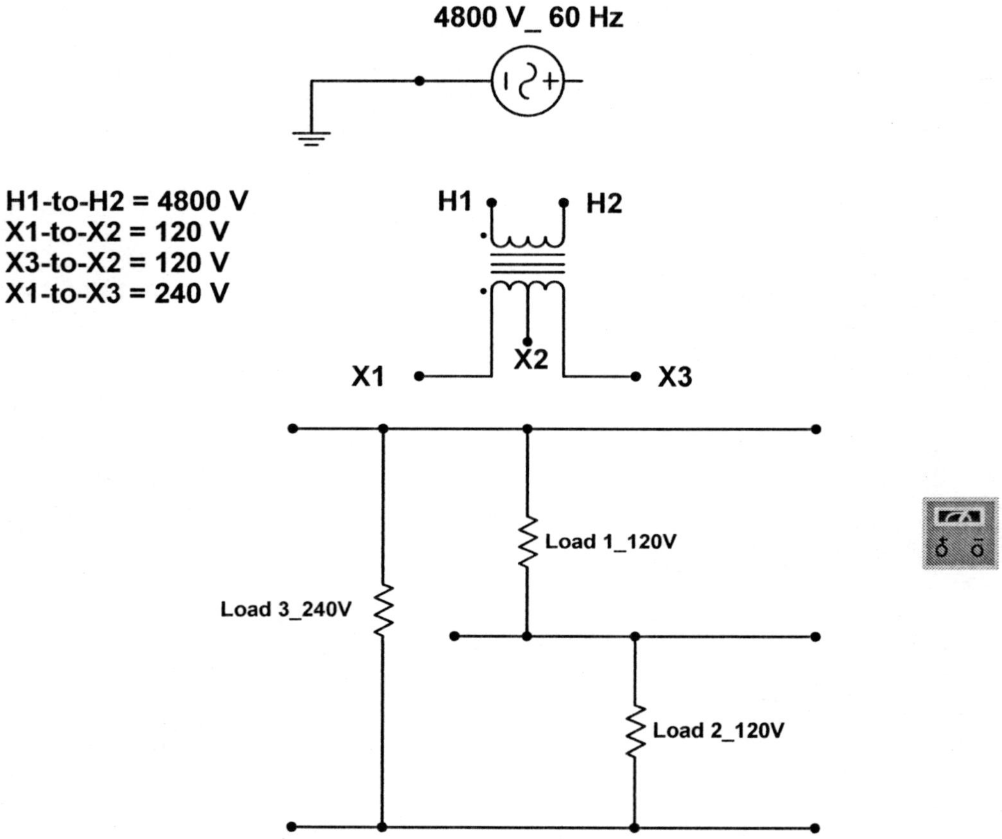

Answers

NAME ___________________________________ DATE __________________

25-8. Open the Multisim circuit file CH25 _08 shown.

(a) Complete the wiring of the transformer so that the two transformers are connected in parallel with each other and supplying power to the load.

(b) Record the value of the voltage across the load.

(c) Record the value of the total current supplied to the load.

(d) Record the value of the current being supplied by the secondary winding of each transformer.

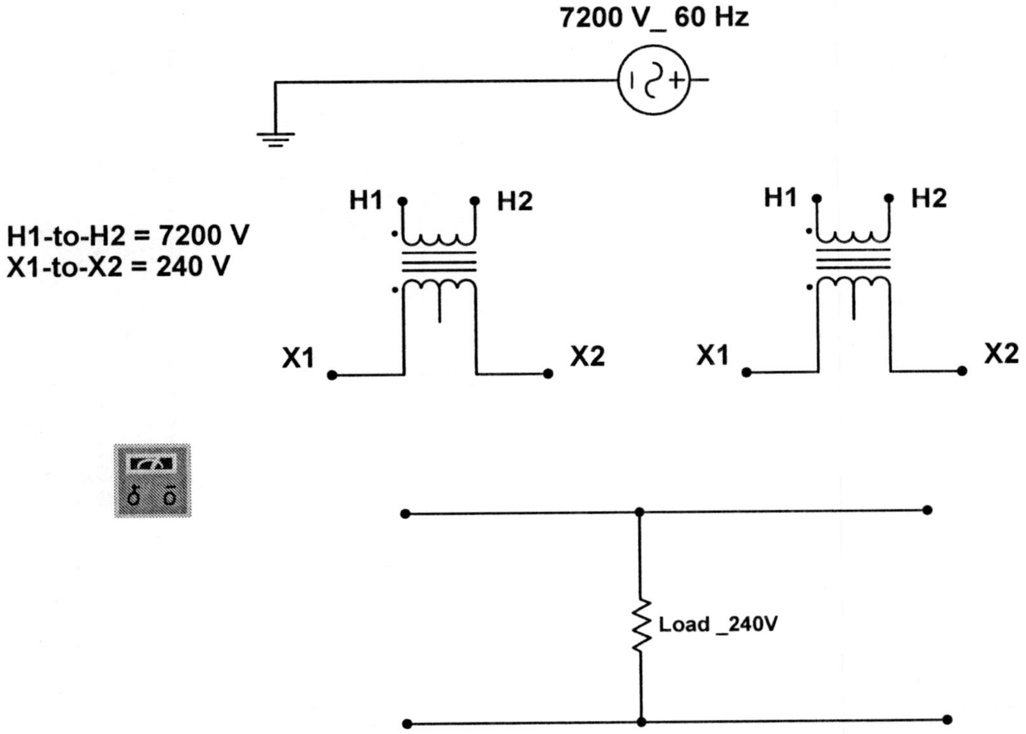

<u>**Answers**</u>

NAME ______________________________ DATE ________________

25-9. Open the Multisim circuit file CH25 _09 shown and turn the simulation on.

(a) Identify the type of 3-phase transformer connection shown.

(b) Record the value of the primary line voltage.

(c) Record the value of the primary phase voltage.

(d) Record the value of the secondary phase voltage.

(e) Record the value of the secondary line voltage.

(f) Calculate the turns ratio for each single phase transformer.

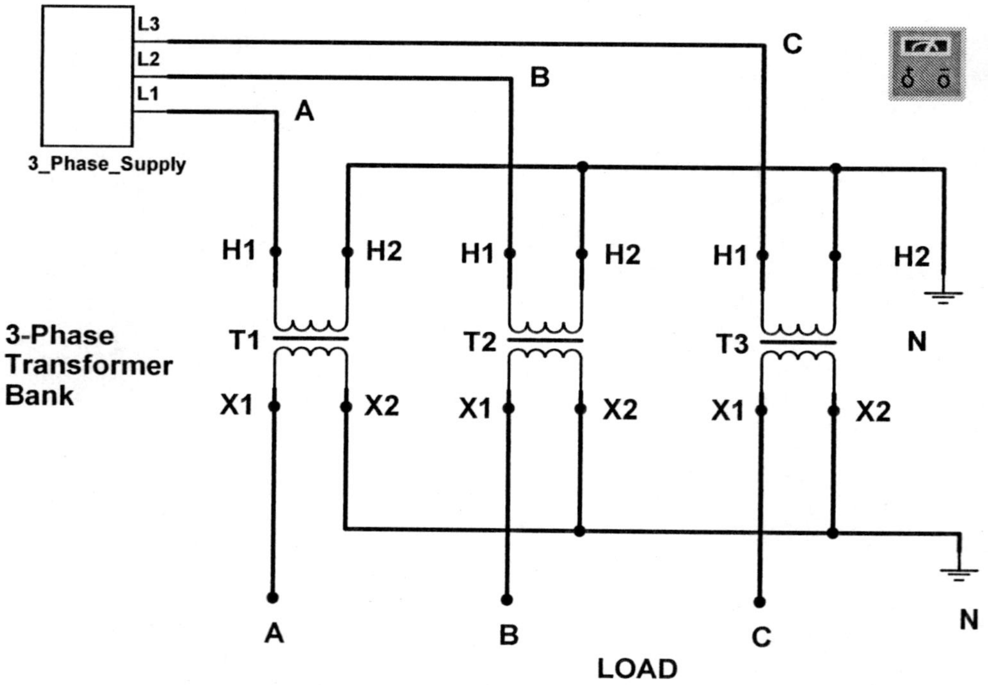

Answers

NAME ______________________________ DATE ________________

25-10. Open the Multisim circuit file CH25 _10 shown and turn the wye-wye transformer simulation on.

(a) Record the value of the primary line voltage as indicated by the voltmeter.

(b) Calculate the value of the primary phase voltage.

(c) Measure the value of the primary phase voltage.

(d) Calculate the value of the secondary phase voltage.

(e) Measure the value of the secondary phase voltage.

(f) Calculate the value of the secondary line voltage.

(g) Measure the value of the secondary line voltage.

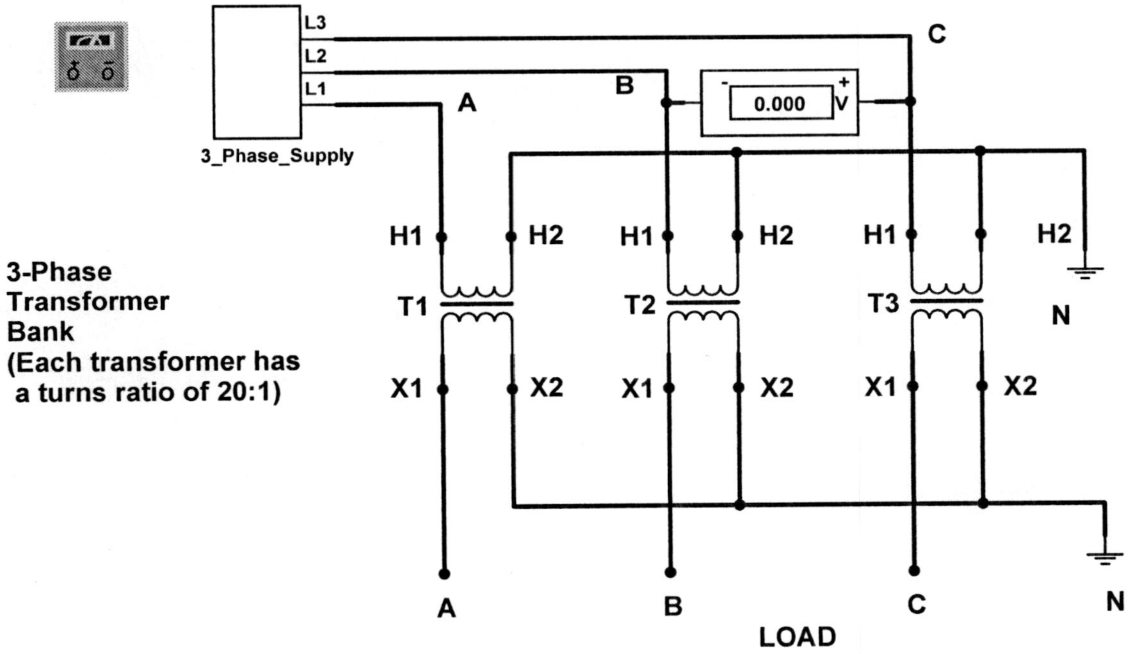

Answers